CONTENTS

QUEUEING THEORY

Worked examples and problems

J

H ...rch Unit,
S
C

M

First published 1978 by
THE MACMILLAN PRESS LTD
London and Basingstoke
Associated companies in New York Dublin
Melbourne Johannesburg and Madras

ISBN 0 333 21702 0

Printed in Hong Kong by Wing King Tong Co., Ltd.

PREFACE

The basic concepts and an understanding of modern queueing theory are requirements not only in the training of operational research staff, management scientists, etc., but also as fundamental concepts in the training of managers or in mangement development programmes.

The efficient design and operation of 'service functions' is one of the main problems facing management today and the understanding obtained from a study of queueing theory is essential in the solution of these problems.

Industry and commerce have for too long concentrated their main resources on designing and operating the 'production units' and little attention has been paid until recently to the 'service units'. Basic concepts such as 'increased efficiency is achieved when the utilisation of service units is reduced' are still hard for practical personnel to understand, brought up as they are on the concept of 'maximising the utilisation' of their facilities. The ancient Chinese civilisation had a system based on queueing theory: 'Pay your doctor only when you are well'. Thus in industry if a system is correctly designed, management should be happy when 'its maintenance gang is playing cards' since there are no breakdowns to be repaired!

This book, by concentrating on problems with their fully worked-out solutions, gives students of queueing theory not only a chance to test their understanding of the theory, but also illustrates the wide range of application of the theory.

The book covers the steady-state solutions of random-arrival queueing systems. It is designed to meet the needs not only of management science training programmes, but also of mangement teaching programmes.

Cranfield, 1976 J. Murdoch

GLOSSARY OF SYMBOLS

D	deterministic distribution
G	general distribution
M	negative exponential distribution
M_n	negative exponential distribution with mean dependent on the number in the system n
C	number of channels
N	maximum system size in finite queues
λ	average arrival rate
$1/\lambda$	average inter-arrival time
μ	average service rate
$1/\mu$	average service time
$\rho = \frac{\lambda}{\mu}$ (or $\frac{\lambda}{C\mu}$)	intensity of traffic for single and multi-channel queues
or $\varepsilon = \lambda/\mu$	traffic offering (multi-channel queues)
σ_s^2	variance of service time
$d(t)dt$	distribution of the time in the system (steady state)
n	number in the system
$\bar{n}$	average number in the system
$P_n(t)$	transient state probabilities of n in the system
P_n	steady-state probabilities of n in the system
q	number in the queue
$\bar{q}$	average number in the queue
$w(t)dt$	distribution of the waiting time in the queue in the steady state
$\bar{w}$	average waiting time of all customers in the queue in the steady state.

BASIC DISTRIBUTIONS

$$P(x) = \frac{e^{-m}m^x}{x!}$$ *Poisson* distribution (mean = m)

$P(t) = \frac{1}{T}e^{-t/T}\,dt$ *Negative exponential* distribution (mean = T)

CLASSIFICATION OF QUEUEING SYSTEMS

Queues are classified in the book as follows.

(1) / (2) / (3) / (4)

(1)	*Input Distribution*	e.g. M, D, M_n, G, etc.
(2)	*Service Distribution*	e.g. G, D, M, etc.
(3)	*Number of Service Channels*	e.g. 1, 2, ... C, etc.
(4)	*Number in the System*	Unconstrained ∞, finite, maximum size = N

EXAMPLES OF USE OF CLASSIFICATION SYSTEM

Thus an $M/M/1/\infty$ system is random arrival, negative exponential service time distribution, single-channel, no constraint on queue size.

Again a G/M/C/N system is general arrival distribution, negative exponential service distribution, C service channels, maximum number in the system N.

The service mechanism in all problems, is service in order of arrival, or first-in, first-out (FIFO) system.

1 BASIC CONCEPTS OF QUEUES

1.1 INTRODUCTION

Queueing situations arise in all aspects of work and life and are typified by the 'queueing for service'. The theory of queueing gives a basis for understanding the various aspects of the problems and enables a quantitative assessment to be made. Therefore the theory enables these 'service situations' to be more effectively designed and operated.

Understanding queueing theory and its concepts is thus basic to all personnel concerned with service situations. Since a large proportion of both capital and labour is tied up in service facilities, and these areas have in the past tended to be neglected for the direct productive units, there is clearly a large potential area of application of the theory and also large savings to be obtained.

This book, by giving a series of problems with their worked solutions, aims not only to teach understanding of the basic theory but also to give readers an insight into the potential of the theory and its wide field of application.

1.2 THE QUEUEING SITUATION

A situation in which queueing can occur may be typified by a shop where customers expect to be served by sales assistants. If all assistants are busy when a new customer enters, he has to wait and thus forms the beginning of a queue. In our discussion of queues in general, we shall call 'customer' the incoming unit, that is, the unit that enters into a situation in which a queue could form; such queues need not take the form of 'customers' actually lining up, all we need, to define a customer as queueing, is the fact that he has made clear his expectation of being served, and that the service is not available. By 'service' we shall mean any action necessary to allow the customer to leave the 'shop', or 'counter' - in general the situation where queueing had been possible. Thus there are three essential elements of any queueing situation: (1) input process - the manner in which customers arrive; (2) queue discipline - the manner in which customers wait for service after input; and (3) service mechanism - the manner in which customers are being served, or the way in which the queue is being resolved.

Figure 1.1 illustrates the queueing system for a shop

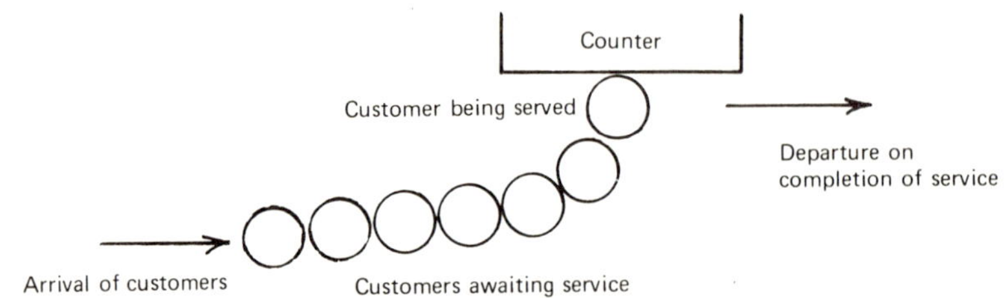

Figure 1.1 Diagrammatic Representation of Single-channel Queue.

with a single server or counter, while figure 1.2 illustrates two different queueing systems for a three-channel system (three counters in parallel).

(a) Single queue

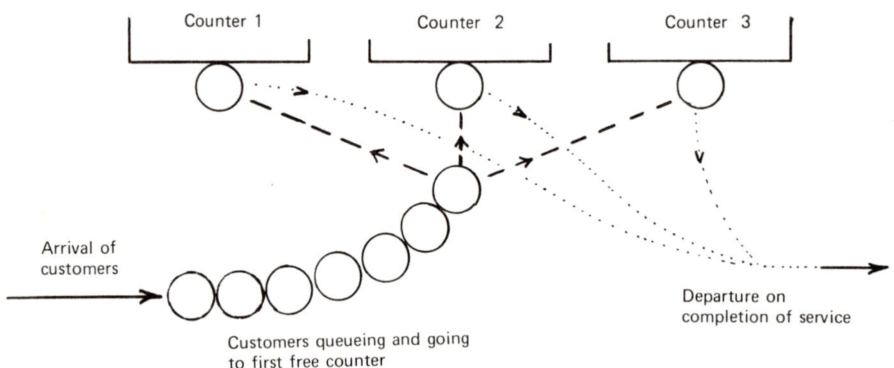

(b) Independent queues

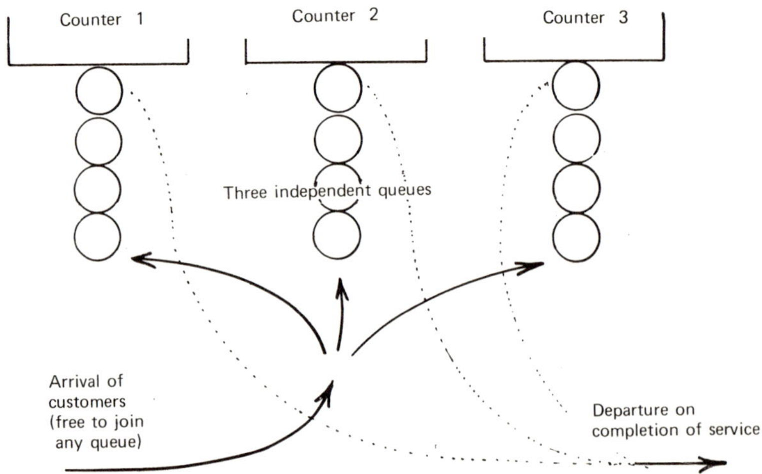

Figure 1.2 Diagrammatic Representation of Three-channel Queue.

Queueing situations are very widespread and many apparently quite different examples can be found in practice. Figure 1.3 shows a few of the more important ones, many of which have been the subject of published investigations.

1.3 TYPES OF QUEUEING PROBLEM

Although all queueing situations are basically similar, there are an almost infinite number of different situations that can arise in practice.

As previously stated, the three basic elements of a queueing problem are

(1) input process
(2) queueing discipline
(3) service mechanism.

These elements have within themselves a large number of possible variations, which give rise to a large number of different queueing situations. Figure 1.4 gives a list of possible variations, although this itself is not exhaustive.

1.4 THE BASIC THEORY

Queueing problems arise, as has been seen, in any activity where demands for service arise from a multiplicity of sources acting more or less independently of each other. Where customer arrivals, or demands, for service can be scheduled exactly, then it is relatively easy to provide appropriate service facilities, and this is really a trivial problem compared with the ones that occur more usually in practice and to which queueing theory give the basis for solution.

In developing the theory, it is convenient to imagine customers arriving at a counter and queueing for service, if the service mechanism is busy (see figures 1.1 and 1.2). This method can also cover situations where a physical queue does not in practice exist, for example, machines awaiting service from an overhead crane, callers waiting on different lines for a connection by the telephone operator, etc.

1.4.1 Measures of Effectiveness

It has been found possible to set up mathematical models to describe queueing situations specified by different forms of the three basic elements. These models can then be manipulated to show what the service system under investigation should be capable of achieving and how any two or more systems compare. In order to make a decision

Figure 1.3 Some Typical Situations for the Application of Queueing Theory

Situation	*Input to Queue*	*Queue*	*Service Mechanism*
Shop Booking office Post Office Bank Hairdressing salon	Customers or clients arriving for service	Waiting for counter to be free	Assistant, teller, etc., serving at counter
Traffic Bus stops Taxi ranks	Customers arriving	Customers queueing waiting for bus or taxi	Arrival of bus, taxi, etc.
Doctor's or hospital outpatients waiting room	Patients arrival for treatment	Patients waiting their turn	Treatment by doctor
Factory handling system	Jobs requiring movements	Jobs waiting at various points for movement	Actual movement of job by transport
Airport	Planes arriving to land	Planes circling overhead waiting for free runway	Planes landing on runway at airport
Stocking of goods	Arrival of batches of goods from supplier	Stocks of goods in store	Usage or purchases of goods from store

Situation	*Input to Queue*	*Queue*	*Service Mechanism*
Telephone switch-board	Customers picking up phone	Customers awaiting telephone switch-board operator's response	Switching at switch-board or exchange
Estimating (tendering)	Possible contracts for pricing for tendering	Contracts awaiting pricing	Pricing by estimator and sending off tender
Harbour design	Ships arriving at port for loading or unloading	Ships awaiting berth	Ships being loaded or unloaded
Semi-skilled oper-ators in machine shop	Operators and/or machine breakage, etc., requiring skilled setter (or maintenance operator)	Operators and machines waiting for skilled setter (or mainten-ance)	Adjustment (or repair) by skilled setter, or maintenance)
Maintenance department	Plant breaks down	Plant awaiting repair	Plant repair

Input Process

Can vary as follows in

1. Number of potential customers
   - (a) Infinite
   - (b) Finite
2. Number arriving at one time
   - (a) Singly
   - (b) In batches of constant number
   - (c) In batches of variable number
3. Intervals between arrivals
   - (a) Constant
   - (b) Completely random (Poisson input)
   - (c) Other distribution of intervals

Queueing Discipline

Can vary as follows in

1. Number of queues
   - (a) Single queue
   - (b) Several queues
2. Queue discipline
   - (a) Service in order of arrival
   - (b) At random
   - (c) Priority
   - (d) Service in reverse order of queueing (unfair queue)
   - (e) Service time-dependent

Service Mechanism

Can vary as follows in

1. Number of service points or servers
   - (a) One
   - (b) Several
   - (c) Number variable
2. Number served
   - (a) One at a time
   - (b) In batches of constant number
   - (c) In batches of variable number
3. Service available
   - (a) Permanently
   - (b) Intermittently

Input Process

4. Average rate of arrival
   - (a) Constant
   - (b) Varying with time
   - (c) Influenced by a state of queue

5. Outside influence
   - (a) No outside influence
   - (b) Input is the output of a previous queueing situation

Queueing Discipline

Service Mechanism

4. Duration of service
   - (a) Constant
   - (b) Exponentially distributed (times of beginning and end of service distributed independently and at random)
   - (c) Other distributions
   - (d) Dependent on time customer has spent in queue

5. Average rate of service
   - (a) Constant
   - (b) Varying with time or state

Figure 1.4 Queueing Systems: Types of Variation in System.

on which system is the 'best', certain 'measures of effectiveness' are required.

Useful measures of effectiveness have been found to be

(1) The probability of having n customers waiting at a time t, given the initial state of the system. Knowing this probability distribution, the size of the queue that will be exceeded for only 5 per cent, say, or 1 per cent of the time, can be determined. This could be useful, for example, in determining what size of waiting room needs to be provided for customers so that only rarely will there be an overflow of customers and possible loss of business if there is an alternative service point they can go to. Alternatively, with a given size of waiting space, the service facility required could be determined such that the waiting space will be adequate most of the time.

(2) The distribution of waiting time of customers. From this distribution can be found the average waiting time of customers and the proportion of customers who have to wait longer than a certain time t, say, If the probability of waiting longer than t is high, then customers may be discouraged from joining the queue, which would result in a loss of potential business in something like a petrol station or a supermarket. The cost of providing more or faster service facilities may be more than compensated for by the extra business produced by a reduction of customers' waiting time. Again, in the case of an internal stores in a factory, provision for an extra storekeeper, say, may pay handsome dividends in reducing the lost production time of skilled men who have to queue for a long time for service.

These measures of effectiveness (depending which, if any, is appropriate to the problem) can be used to decide, for instance (usually on a cost basis) whether to speed up the existing service rate of each channel or whether to provide extra channels working at the same rate as the present ones or whether even a reduction in service facility can be contemplated.

1.5 MATHEMATICAL SOLUTION OF QUEUEING PROBLEMS

1.5.1 Transient and Steady-state Solutions

The solution of queueing problems is considered in two parts, namely the transient or time-dependent solution, and the steady-state solution. Briefly, provided that the service channel is capable of serving at a faster average rate than that at which customers arrive, then the steady-state is reached when the queue behaves independently of the initial state of the system, and the probability of having a given number, n, say, in the queue remains constant with time. This situation exists, more or less, in a machine shop where the queue of demands for the overhead crane unsatisfied at the end of the day is carried over to the next day (assuming that the crane only operates during normal working hours and does not work off the backlog of jobs during the night, say). The state of the queue soon becomes independent of the starting conditions when the present pattern of production was begun.

On the other hand, in somewhere like a bank, the system starts every day with no people at all either being served or queueing, and the chance of finding, say, 6 people queueing depends on how soon after opening time the observation is made. Assuming that there are always the same number of clerks on duty and customers arrive at a constant average rate throughout the day, the chance of finding 6 people queueing immediately after the bank opens is likely to be very small indeed. As the day proceeds, the queue gradually achieves a steady state and eventually the queue fluctuates about a fixed average size, the probability of finding 6 people queueing now being higher than it was at the very start of the day's business.

In most practical situations only the steady-state queue need be considered, but occasionally only the transient solution is applicable since the queue never reaches a steady state. This latter is generally true if the arrival rate is greater than the service rate ($\lambda > \mu$) and applies in some cases when the system is not in operation long enough before reverting to the starting state, usually with no customers at all in the system.

This book deals only with the steady-state solution results, and the use of the theory is demonstrated with a range of problems for each system. The queueing systems covered relate only to random-arrival systems.

2 BASIC DISTRIBUTIONS IN QUEUEING THEORY

2.1 INTRODUCTION

The Poisson distribution and the negative exponential distribution are the two basic distributions in queueing theory. Their theory and general fields of application can be studied in general statistical theory texts and this book will deal primarily with their application to queueing theory.

2.2 RÉSUMÉ OF BASIC THEORY AND FORMULAE

2.2.1 Poisson Distribution

General Law

If the chance of an event occurring at any instant of time is constant in a continuum of time and if the average number of successes in time t is m, then the probability of x successes in time t is

$$P(x) = \frac{m^x e^{-m}}{x!}$$

with mean of distribution = m
and variance of distribution = m

The Poisson distribution is tabulated in table 1, in the statistical tables at the end of this book.

2.2.2 Negative Exponential Distribution

General Law

If the chance of an event occurring at any instant of time is constant in a continuum of time, then if the average time interval between successes is T, then the probability of an interval t between successes is

$$P(t) = \frac{1}{T} e^{-t/T} dt$$

Thus both these distributions describe the same random situation. Also, the probability of a time interval exceeding t is

$$= \int_t^\infty \frac{1}{T} e^{-t/T} dt$$

$$= e^{-t/T}$$

This function is tabulated in table 2 of the statistical tables at the end of this book.

2.2.3 Special Property of the Negative Exponential Distribution

The negative exponential distribution of service times is important in queueing theory because of the following special property, namely that the time a service has been in progress does not affect the probability of its completion. This proof is given below.

Let service have been in progress for time t. Then it must be 'at least' time t long, the probability of this being

$$P(>t) = \int_t^\infty \frac{1}{T} e^{-t/T} dt$$

$$= e^{-t/T}$$

If the service now ends during interval dt, it must therefore have had time between t and (t + dt), this probability being

$$P(t)dt = \frac{dt}{T} e^{-t/T}$$

Now P(t)dt = (probability of call lasting till time t) × conditional probability that it does not last beyond (t + dt)

Therefore given that the call has lasted till t

$$\text{probability that it finishes in interval (t) to (t + dt)} = \frac{P(t)dt}{P(>t)}$$

$$= \frac{(dt/T)e^{-t/T}}{e^{-t/T}}$$

$$= \frac{1}{T} dt$$

or this probability is independent of length of time (t) the call has been in progress.

2.3 PROBLEMS

1. Customers arrive at a store for service at an average rate of 10 per hour. Given that customers arrive randomly, what is the probability of

(a) more than 15 customers arriving in one hour?
(b) exactly 10 customers arriving in one hour?
(c) more than 6 customers arriving in half an hour?

2. Draw the distribution of number of arrivals per hour given that average number of arrivals per hour = 3, and that customers arrive randomly.

3. Customers arrive randomly at a service point with an average rate of 4 per hour. Draw the distribution of the interval between successive arrivals.

4. The time taken to repair a machine is distributed as the negative exponential distribution with a mean of 10 hours. What is the probability that

(a) a machine takes longer than 6.9 hours to repair?

(b) a machine takes longer than 10 hours to repair?

What is the repair time that is exceeded by chance once in 100 repairs?

5. The failure rate for a television receiver is 0.02 failures per hour. Calculate the average time between failures. What is the probability of it failing within 4 hours?

6. The time interval, in minutes, between the arrival of successive customers at a cash desk of a self-service store was measured over 56 customers and the results are given below.

Time Interval between Arrivals (min)						
0.05	1.68	0.78	1.10	0.32	1.61	0.10
0.21	2.71	2.12	2.81	3.30	0.15	0.54
1.14	0.16	0.31	0.91	0.18	0.04	1.16
0.57	0.65	4.60	1.72	0.52	2.32	0.08
1.16	0.58	0.57	0.04	1.19	0.11	0.05
0.15	0.42	0.25	0.05	1.18	3.90	0.01
0.43	2.16	2.68	0.08	4.20	0.09	0.63
3.12	0.62	3.70	1.48	2.08	1.76	1.21

Fit a negative exponential distribution to the data.

2.4 SOLUTIONS

1(a) Here average arrival rate per hour m = 10, therefore

$$\text{Probability of more than 15 customers in one hour} = \sum_{x=16}^{\infty} \frac{e^{-10}10^x}{x!}$$

From table 1

$P(>15) = 0.0487$

(b) Again from table 1

$P(10) = 0.5421 - 0.4170 = 0.1251$

(c) From table 1, now m = 5

$P(>6) = 0.2378$

2. From table 1

Prob. of no customers arriving $P(0) = 0.0492$

$P(1) = 0.1493$
$P(2) = 0.2241$
$P(3) = 0.2240$
$P(4) = 0.1681$
$P(5) = 0.1008$
$P(6) = 0.0504$
$P(7) = 0.0216$
$P(>7) = 0.0119$

Figure 2.1 shows the histogram.

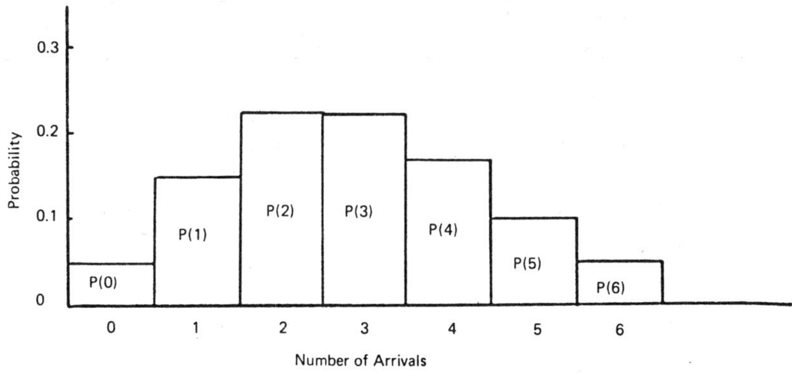

Figure 2.1

3. Here the average interval between arrivals = 0.25 hours. Distribution is

$$P(t) = \frac{1}{0.25} e^{-t/0.25} dt$$

All probability of exceeding t

$$= \int_t^{\infty} \frac{1}{0.25} e^{-t/0.25} dt$$

$$= e^{-t/0.25}$$

This exponential function is tabulated in table 2.

Interval between Arrivals h (t)	*Probability of Exceeding t P(>t)*
0.00	1.0000
0.05	0.8187
0.10	0.6703
0.15	0.5488
0.20	0.4493
0.30	0.3012
0.40	0.2019
0.50	0.1353
0.75	0.0498
1.00	0.0183

Therefore the probability distribution is as follows.

Interval between Arrivals (h)	*Probability*
0.00 - 0.05	0.1813
0.05 - 0.10	0.1484
0.10 - 0.15	0.1215
0.15 - 0.20	0.0995
0.20 - 0.30	0.1481
0.30 - 0.40	0.0993
0.40 - 0.50	0.0666
0.50 - 0.75	0.0855
0.75 - 1.00	0.0315
over 1.00	0.0183
Total	1.0000

This probability distribution is drawn in figure 2.2.

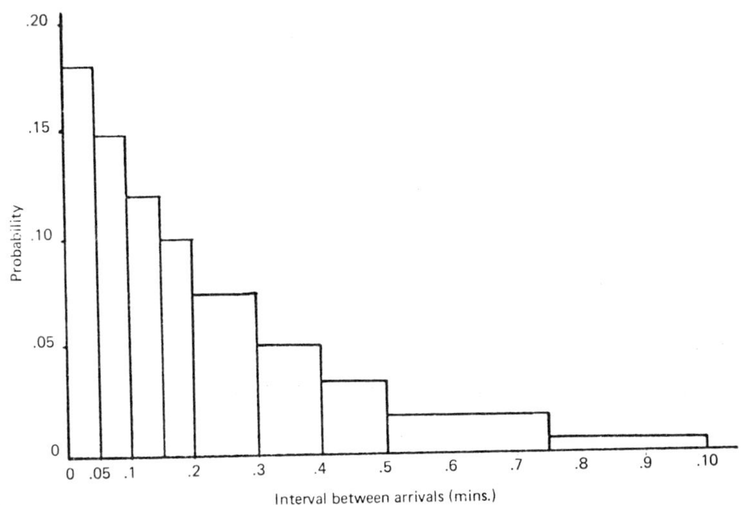

Figure 2.2

4. Here T = 10 hours. From table 2

(a) Probability of repair exceeding 6.9 hours $= e^{-6.9/10} = e^{-0.69}$

$= 0.5016$

(b) Probability that repair exceeds 10 hours $= e^{-10/10} = e^{-1}$

$= 0.3679$

(c) From table 2, the value of m in the function e^{-m} that gives a value of 0.01 to the function is m = 4.61, therefore

$$\frac{t}{T} = 4.61 \quad \text{and} \quad t = 46.1 \text{ h}$$

or a repair time of 46.1 h or more is to be expected once in 100 times on the average.

5. If the failure rate is 0.02 failures per hour, then the average time between failures

$$= \frac{1}{0.02} = 50 \text{ h}$$

Probability of failing inside 4 h = 1 - Probability of failing in over 4 h

$$= 1 - e^{-4/50}$$

$$= 1 - e^{-0.08}$$

from table 2

$$= 1 - 0.9231$$

$$= 0.0769$$

6. The data are summarised into a distribution in the following table, together with the negative exponential distribution

Fitting a negative exponential distribution

Average interval between arrivals

$$T = \frac{19 \times 0.25 + 11 \times 0.75 + \ldots + 1 \times 4.25 + 1 \times 4.75}{56}$$

$$= \frac{72.5}{56} = 1.29$$

From Table 2

$$P(t > 0.50 \text{ min}) = e^{-0.5/1.29} = e^{-0.39} = 0.6771$$

therefore

Probability of interval between 0 and 0.50 min = 1 - 0.6771 = 0.3229

Again from Table 2

$$P(t > 1.00 \text{ min}) = e^{-1.00/1.29} = e^{-0.78} = 0.4584$$

therefore

Probability of interval between 0.50 and 1.00 min = 0.6771 - 0.4584

= 0.2187

etc.

The fitting of the negative exponential distribution is summarised in the following table.

	ACTUAL		NEGATIVE EXPONENTIAL	
	Frequency	*Probability*	*Probability*	*Frequency*
0.00 - 0.49	19	0.340	0.3229	18.1
0.50 - 0.999	11	0.196	0.2187	12.2
1.00 - 1.499	7	0.125	0.1449	8.1
1.50 - 1.999	6	0.107	0.1013	5.7
2.00 - 2.499	4	0.071	0.0685	3.8
2.50 - 2.999	3	0.054	0.0464	2.6
3.00 - 3.499	2	0.036	0.0308	1.7
3.50 - 3.999	2	0.036	0.0215	1.2
4.00 - 4.499	1	0.018	0.0145	0.8
4.50 - 4.999 (and over)	1	0.018	0.0305	1.7
	56	1.000	1.000	55.9

3 M/M/1/ SYSTEMS

3.1 INTRODUCTION

The theory of this system and also the other systems described in later chapters can be found in most text-books on queueing theory and a list of references is given at the end of this book.

In the M/M/1/∞ system, customers arrive randomly for service - Poisson stream, service time distribution is negative exponential, single server, no constraint on queue size. Customers are served in order of arrival.

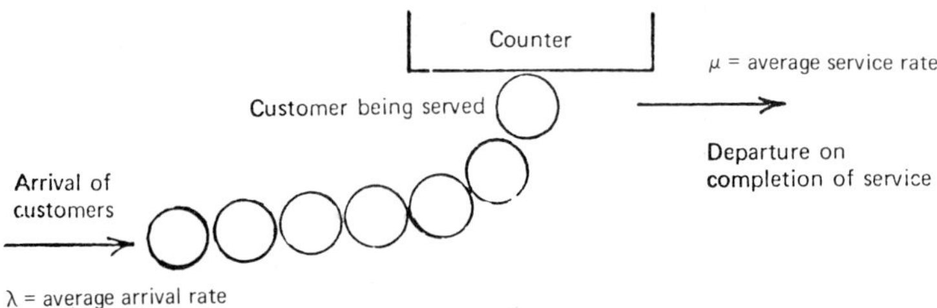

Figure 3.1 M/M/1/∞ Systems

3.2 RÉSUMÉ OF BASIC THEORY AND FORMULAE

The basic formulae of these systems are given below.

Probability of no customers in the system $P_0 = 1 - \rho$

Probability of n customers in the system $P_n = (1 - \rho)\rho^n$

Average number of customers in the system $\bar{n} = \frac{\rho}{1 - \rho}$

Average queue length $q = \frac{\rho^2}{1 - \rho}$

Probability that there are more than r customers in the system $P(>r) = \rho^{r+1}$

Probability that there are more than r customers in the queue $= \rho^{r+2}$

Waiting Time

Waiting time distribution $w(t) = \rho(\mu - \lambda)e^{-t(\mu-\lambda)}dt$

Average waiting time $\bar{w} = \frac{\rho}{1 - \rho} \times \frac{1}{\mu}$

Probability of waiting time greater than t $= \rho e^{-t(\mu-\lambda)} dt$

Distribution of total time in the system $d(t) = (\mu - \lambda)e^{-t(\mu-\lambda)}$

Average time in the system $\bar{d} = \frac{1}{\mu - \lambda} = \frac{1}{\lambda} \cdot \frac{1}{1-\rho}$

Probability of spending longer than t in the system $= e^{-t(\mu-\lambda)}$

3.3 PROBLEMS

1. In a single-channel queueing situation, an activity-sampling analysis of the system gave the following details of the number of persons in the system.

No. of Persons	*Times Observed*
0	900
1	821
2	658
3	621
4	503
5	420
6	302
7	252
8	181
9	152

Note: The data for more than 9 in the system is not included.

Do these results support the hypothesis that the queueing situation is an $M/M/1/\infty$ system?

2. A television repairman finds that the time spent on his jobs has an exponential distribution with mean 30 minutes. If he repairs sets in the order in which they come in, and if the arrival of sets is distributed as Poisson with an average rate of 10 per 8-hour day, what is the repairman's expected idle time each day?

What is the average number of sets ahead of a set that has just been brought in?

3. In the design of the layout of handling equipment for an unloading bay at a factory, three schemes (A, B, C) are being considered, relevant details being as follows.

Scheme	*Fixed Cost /Day* (£)	*Variable Op. Cost/Day* ($£c_v$)	*Handling Rate/Hour* (No.of sacks)
A	60	100	1000
B	130	150	2000
C	250	200	6000

Average arrival rate of trucks = 15 per 10-hour day; average truck load = 500 sacks.

If the cost of a truck waiting is given to be £10 per hour and the queueing is $M/M/1/\infty$, which scheme gives overall minimum cost?

4. In a supermarket the average arrival rate of customers is 5 every 30 minutes. The average time it takes to list and calculate the customer's purchases at the cash desk is 4.5 minutes, and this time is exponentially distributed.

(a) How long will the customer expect to wait for service at the cash desk?

(b) What is the chance that the queue length will exceed 5?

(c) What is the probability that the cashier is working?

5. In question 4, if by the application of work study techniques the average service time is reduced to 4 minutes, how long will customers have to wait on average under this system?

What is the probability that the customer will have to wait more than 10 minutes for service?

6. At what average rate must a clerk at a supermarket work in order to ensure a probability of 0.90 that the customers will not have to wait longer than 12 minutes? It is assumed that there is only one counter, at which the customers arrive in a Poisson fashion, at an average rate of 15 per hour. The length of service by the clerk has an exponential distribution.

7. During certain weekdays the average arrival of customers for service at the deposit counter in a bank is 4 per hour. The average time to complete the deposit is 6 minutes with a negative exponential distribution. Calculate the average queue size and the probability of having more than 4 persons in the queue.

At Friday lunch time the arrival rate goes up to 8

per hour. What must the service time be reduced to in order to ensure only a 1 in 100 chance of any person queueing for more than 12 minutes?

8. A repairman is to be hired to repair machines that break down at an average rate of 3 per hour. Breakdowns are distributed in time in a manner that may be regarded as Poisson. Non-productive time on any one machine is considered to cost the company £5 per hour. The company has narrowed the choice down to two repairmen, one slow but cheap, the other fast but expensive. The slow cheap repairman asks £3 per hour; in return he will service breakdown machines exponentially at an average rate of 4 per hour. The fast expensive repairman demands £5 per hour, and will repair machines exponentially at an average rate of 6 per hour. Which repairman should be hired?

9. A company is considering installing a tool-grinding machine for use by its operators. The following proposals are under review.

Machine A cost £2000 Average grinding time = 10 minutes (negative exponential distribution)

Machine B cost £6000 Average grinding time = 8 minutes (negative exponential distribution)

Machine C cost £10000 Average grinding time = 4 minutes

(All machines have to be written off in 2 years' time).

If the demand rate of operators for grinding is 5 per hour, and the cost of the operators' non-productive time (including wages and lost production) is £2 per hour, which machine should be installed by the company? (Assume 50 weeks/year, 40 hours/week.)

10. Ships arrive randomly at a harbour, and the unloading time is 1 day on the average. It is negatively exponentially distributed. Given a 5-day working week, calculate the distribution of ships' waiting time for

(a) average arrival of 3 ships/week

(b) average arrival of 4 ships/week.

11. In a supermarket, the company's policy is that customers should wait only an average of 2 minutes for service. Given that customers arrive randomly at an average rate of twenty per hour and the average service time is 2.2 minutes (negatively exponentially distributed), what will be the actual average waiting time?

By how much must the average service time be reduced to give an average waiting time of 2 minutes?

12. Patients arrive at the casualty department of a hospital at random with an average arrival rate of 3 per

hour. The department is served by one doctor who spends on average 15 minutes with each patient, actual consulting times being exponentially distributed.

(a) What proportion of the time is the doctor idle (that is, has no patients to examine)?

(b) How many patients are, on average, waiting to see the doctor?

(c) What is the probability of there being more than 3 patients waiting to see the doctor?

(d) What is the average waiting time of patients?

(e) What is the probability of a patient having to wait longer than one hour?

13. Before parts are assembled into a vacuum tube, they must be cleaned in a degreaser. Batches of parts are brought in randomly at an average rate of λ batches per hour, and are cleaned at an average rate of μ batches per hour. The cost of delay is £C_1 per batch per hour, and the cost of owning and operating a degreaser that works at an average rate of μ is £μC_2 per hour. Prove that at minimum total cost

$$\mu = \lambda + \sqrt{\frac{C_1\lambda}{C_2}}$$

3.4 SOLUTIONS

1. For an M/M/1/∞ system, the probability that there are n persons in the system is

$$P_n = \rho^n(1 - \rho)$$

Taking logarithms

$$\log P_n = n \log \rho + \log (1 - \rho)$$

Therefore, if $\log P_n$ is plotted against n, the points should fall on a straight line with slope $\log \rho$ and intercept $\log (1 - \rho)$. Now

$$P_n = \frac{f_n}{\sum_i f_i}$$

Thus $\log P_n = \log f_n - \log \Sigma f_i$

therefore

$$\log f_n = n \log \rho + \log (1 - \rho) + \log \Sigma f_i$$

Therefore, if $\log f_n$ is plotted against n, the points should fall on a straight line with slope log ρ and intercept log $(1 - \rho) + \log \Sigma f_i$.

No. of Persons in System (n)	*Frequency (f_n)*	*Log f_n*
0	900	2.9542
1	821	2.9143
2	658	2.8182
3	621	2.7931
4	505	2.7033
5	420	2.6232
6	302	2.4800
7	252	2.4041
8	181	2.2577
9	152	2.1818
Total	4810	

Plot log f_n against n (see figure 3.2). For accuracy, a regression line should be fitted to the points, but for the purpose of this question it is sufficient to fit a line by visual inspection. The scatter of the points about this line is small, and it is safe to assume that within the error of sampling, these points fall on a straight line. Therefore the survey does indicate that the queueing process conforms to an M/M/1/∞ system.

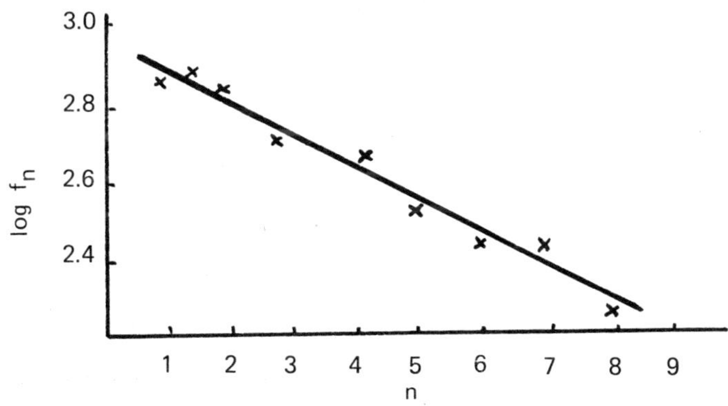

Figure 3.2

Estimation of Traffic Intensity (ρ)

From the graph, the slope is measured as -0.0833, that is $\bar{1}.9167$. Thus $\log \rho = \bar{1}.9167$

traffic intensity $\rho = 0.825$

2. Arrival rate λ = 10 sets per day

Service rate μ = 16 sets per day

$$\rho = \frac{\lambda}{\mu} = \frac{5}{8}$$

Expected idle time per day = proportion of time system empty

$$P_0 = 1 - \rho = \frac{3}{8}$$

that is, 3 hours per day, on average, the repairman is idle. Average number of television sets in the system is

$$\bar{n} = \frac{\rho}{1 - \rho} = \frac{5/8}{3/8} = \frac{5}{3}$$

Therefore average number of sets ahead of new arrival = 5/3.

3. Here average arrival rate λ = 1.5 trucks per hour. The service rate μ depends on the system. Average unloading time per truck is

$$\bar{S} = \frac{\text{no. of sacks/truck}}{\text{unloading rate/h}} \quad \text{(h)}$$

Average time a truck is in the system is

$$\bar{d} = \frac{1}{\mu - \lambda} \quad \text{(h)}$$

therefore

$$\text{delay cost/truck arrival} = \frac{1}{\mu - \lambda} \times £10$$

and $$\text{delay cost/day} = \frac{1}{\mu - \lambda} \times £10 \times 15$$

Actual variable operating cost/day = variable cost/day × utilisation of system $= £\, C_\nu \times \rho$

Scheme	*Average Unloading Rate/Hour (μ)*	*Average Arrival Rate/Hour (λ)*	*Utilisation (ρ)*	*Average Time in System $\bar{d}$ (h)*
A	2	1.5	0.75	2.0
B	4	1.5	0.375	0.40
C	12	1.5	0.125	0.095

The costs of each system are given below.

Scheme	Fixed Cost/day (£)	Variable Cost/day (£)	Delay Cost/day (£)	Total Cost/day (£)
A	60	75	300	435
B	130	56	60	246
C	250	25	14	289

Thus scheme B has the lowest total cost.

4. Assume random arrivals

λ = 5 every 30 min = $\frac{1}{6}$ per min

Average service time = $\frac{1}{\mu}$ = $\frac{9}{2}$ min

thus $\mu = \frac{2}{9}$ per min and $\rho = \frac{\lambda}{\mu} = \frac{1}{6} \times \frac{9}{2} = \frac{3}{4}$

(a) Average waiting time $\overline{w} = \frac{\lambda}{(\mu - \lambda)\mu}$ min

$$= \frac{\frac{1}{6}}{\left(\frac{2}{9} - \frac{1}{6}\right)\frac{2}{9}} = 13.5 \text{ min}$$

(b) Probability of n customers in the system = P_n

Probability of n customers queueing = P_{n+1}

Probability of more than n customers queueing $= \sum_{i=n+1}^{\infty} P_{i+1}$

$= \rho^{n+2}$

therefore

Probability of more than 5 customers queueing $= \frac{3}{4}^{7} = 0.133$

(c) Probability that cashier is working = probability of one or more customers in the system = 1 - probability of no customers in system

$$= 1 - P_0$$

also

$$P_0 = 1 - \rho$$

thus probability that cashier is working $= \rho = \frac{3}{4} = 0.75$

5.

$$\lambda = \frac{1}{6}/\text{min} \qquad \text{and } \rho = \frac{2}{3}$$

$$\mu = \frac{1}{4}/\text{min}$$

$$\text{Average waiting time } \bar{w} = \frac{\frac{1}{6}}{\left(\frac{1}{4} - \frac{1}{6}\right)\frac{1}{4}} \text{ min}$$

$$= 8 \text{ min}$$

Waiting time distribution

$$w(t)dt = \rho(\mu - \lambda)e^{-t(\mu-\lambda)}dt$$

$$\text{Probability of waiting longer than T minutes} = \int_{t=T}^{\infty} w(t)dt$$

$$= \rho e^{-T(\mu-\lambda)}$$

with $T = 10$ min

therefore

$$\text{Probability of waiting longer than 10 minutes} = \frac{2}{3}e^{-5/6} = \frac{2}{3} \times 0.435$$

$$= 0.29$$

6. Let service rate of clerk $= \mu$ per hour. Then

$$\text{probability of waiting longer than T} = \int_{T}^{\infty} w(t)dt$$

$$= \frac{\lambda}{\mu}e^{-(\mu-\lambda)T}$$

Given $\lambda = 15$ per hour

$T = 0.20$ h

therefore

$$\text{probability of waiting longer than 12 minutes} = \frac{15}{\mu}e^{-(\mu-15)0.20}$$

This expression must be less than or equal to 0.1, therefore

$$0.1 \geq \frac{15}{\mu} e^{-(\mu-15)0.20}$$

Giving μ = 25 per hour. (μ = 24 gives 0.1033, 25 correct to nearest whole number.)

7. Average queue length is

$$\bar{q} = \sum_{n=1}^{\infty} (n - 1)P_n = \frac{\rho^2}{1 - \rho}$$

with λ = 4/h and μ = 10/h

$$\bar{q} = \frac{\frac{2}{5}^2}{1 - \frac{2}{5}} = 0.267$$

Probability of more than 4 in queue $= \sum_{n=6}^{\infty} P_n$

$= \rho^6 = 0.004$

Now λ = 8/h

Probability of waiting time $> \frac{1}{5}$ h $= \int_{1/5}^{\infty} w(t)dt$

$= \rho e^{-(\mu-\lambda)/5}$

This must be less than or equal to 0.01, that is

$$\log_e \rho e^{-(\mu-\lambda)/5} < \log_e 0.01$$

On substituting various values for μ, the value (to the nearest whole number) μ = 26/h satisfies the above inequality.

8. Breakdowns random, λ = 3 per hour.

Slow Man

Repairs machines exponentially μ = 4 per hour. Average waiting time of a broken machine is

$$\bar{w} = \frac{\lambda}{(\mu-\lambda)\mu} = \frac{3}{(4-3)4}$$

$$= \frac{3}{4}\text{ h}$$

Average service time = 1/4 h, therefore

Average time to complete repair = 1 h

therefore

cost per hour of machine breakdowns = 1 × 5 × £3 = £15
labour cost per hour = £3
total cost per hour = £18

Fast Man

μ = 6 per hour. Average waiting time is

$$\bar{w} = \frac{3}{(6-3)6} = \frac{1}{6}\text{ h}$$

Average service time = 1/6 h and average time machine is out of action = 1/3 h, therefore

cost per hour = £$\frac{1}{3}$ × 5 × 3 = £5

Labour cost per hour = £5

total cost per hour = £10

Hiring the fast man therefore gives a lower cost per hour.

9. Consider a period of 1 hour.

Machine A

$$\text{Capital depreciation/hour} = \frac{£2000}{2 \times 50 \times 40} = £0.5/\text{h}$$

Machine B

$$\text{Capital depreciation/hour} = \frac{£6000}{2 \times 50 \times 40} = £1.5/\text{h}$$

Machine C

$$\text{Capital depreciation/hour} = \frac{£10000}{2 \times 50 \times 40} = £2.5/\text{h}$$

Consider machine A

Average arrival rate λ = 5/h

Average service rate μ = 6/h

thus $\rho = \frac{5}{6}$

Utilisation of grinding machine = 5/6 = 0.83 or 83%.

Average delay per call (including service time) $= \frac{1}{\mu - \lambda} = \frac{1}{6 - 5} = 1$ h

Total cost of lost time/hour = 5 × 1 × £2 = £10/h

Consider machine B

Average arrival rate λ = 5/h

$\mu = \frac{60}{8} = 7.5$/h

thus $\rho = \frac{5}{7.5} = 0.67$

Utilisation of grinding machine = 67%.

Average delay per call (including service time) $= \frac{1}{\mu - \lambda} = \frac{1}{2.5} = 0.40$ h

Total cost of lost time/hour = 5 × 0.40 × £2 = £4

Consider machine C

Average arrival rate λ = 5/h

μ = 15/h

thus $\rho = \frac{5}{15} = 0.33$

Utilisation of grinding machine = 33%.

Average delay per call $= \frac{1}{\mu - \lambda} = \frac{1}{15 - 5} = 0.1$ h

Total cost of lost time/hour = 5 × 0.1 × £2 = £1/h

Summary

	Total Cost (£)		
	Machine A	*Machine B*	*Machine C*
Capital depreciation	0.5	1.5	2.5
Cost of lost time	10	4	1
Total	10.5	5.5	3.5

Thus machine C (the most expensive) is the best installation, although, as will be seen, all the machines have the capacity to handle the service.

10. Random arrival and exponential service time mean the waiting-time distribution is exponential.

$$w(t)dt = \rho(\mu - \lambda)e^{-t(\mu-\lambda)}dt$$

with mean

$$\bar{w} = \frac{\rho}{(1 - \rho)\mu}$$

(a) Average arrival of 3 ships/week, therefore $\lambda = 3/5$ per day, $\mu = 1$ per day, therefore $\rho = 3/5$.

$$w(t)dt = \frac{3}{5}\left(1 - \frac{3}{5}\right)e^{-t(1-3/5)}dt$$

$$= \frac{6}{25}e^{-(2/5)t}dt$$

$$\text{Mean waiting time } \bar{w} = \frac{\frac{3}{5}}{\left(1 - \frac{3}{5}\right)1} \text{ days}$$

$$= \frac{3}{5} \times \frac{5}{2} = 1.5 \text{ days}$$

(b) Average arrival of 4 ships/week, therefore $\lambda = 4/5$ per day, $\mu = 1$ per day, therefore $\rho = \lambda/\mu = 4/5$. Waiting time distribution is

$$\bar{w}(t) = \frac{4}{5}\left(1 - \frac{4}{5}\right) e^{-t(1-4/5)}dt$$

$$= \frac{4}{25} e^{-t/5}dt$$

$$\text{Mean waiting time } \bar{w} = \frac{\frac{4}{5}}{\left(1 - \frac{4}{5}\right)1} = 4 \text{ days}$$

11. Random arrival: $\lambda = 20/\text{h}$; exponential service time: $\mu = (60/2.2)/\text{h}$, therefore $\mu = 27.3/\text{h}$.

(a) Average waiting time $\bar{w} = \dfrac{\rho}{(1 - \rho)\mu}$

$$= \frac{20/27.3}{(1 - 20/27.3)27.3}$$

$$= \frac{0.733}{0.267 \times 27.3}$$

$$= 6.0 \text{ min}$$

(b) $\bar{w} = \frac{\rho}{(1 - \rho)\mu} = 2 \text{ min}$

that is

$$= \frac{\lambda/\mu}{\left(1 - \frac{\lambda}{\mu}\right)\mu} = 2$$

thus $2\mu^2 - 2\mu\lambda - \lambda = 0$

$$2\mu^2 - \frac{2}{3}\mu - \frac{1}{3} = 0$$

$$6\mu^2 - 2\mu - 1 = 0$$

$$\mu = \frac{2 \pm \sqrt{(14 + 24)}}{12} \text{ per min}$$

One root is negative, thus $\mu = (2 + 5.3)/12 = 0.61$ per minute, that is 60×0.61 customers per hour must be served.

average service rate $\mu = 36.6$ per hour

Thus average service time $= 60/36.6 = 1.64$ min.

12.

(a) Probability (doctor idle) = probability (system empty)

$= 1 - \rho$

$$\rho = \frac{\lambda}{\mu} = \frac{3}{4}$$

therefore doctor is idle for 25 per cent of the time.

(b) Average number of patients waiting to see the doctor = average number in queue

$$\bar{q} = \sum_{r=1}^{\infty} (r - 1)P_r = \frac{\rho^2}{1 - \rho}$$

thus average number waiting $= \frac{(3/4)^2}{1 - \frac{3}{4}} = 2.25$

(c) Probability (more than 3 in queue) = Probability (more than 4 in system)

$$= \sum_{r=5}^{\infty} P_r = \rho^5 = \left(\frac{3}{4}\right)^5$$

$$= 0.24$$

(d) Average waiting time $\bar{w} = \frac{\rho}{(1 - \rho)\mu}$

$$= \frac{3/4}{1 - \frac{3}{4}4} \text{ h}$$

$$= 0.75 \text{ h}$$

(e) Probability of a patient having to wait longer than one hour

$$= \int_1^\infty w(t)\,dt$$

$$= \rho e^{-(\lambda - \mu)} = \frac{3}{4} e^{-(4-3)}$$

$$= \frac{3}{4} \times 0.3679 = 0.276$$

13.

Average time batch spends in system $= \bar{d} = \frac{1}{\mu - \lambda}$ h

therefore

delay cost per batch $= £\frac{C_1}{\mu - \lambda}$

delay cost per hour $C_D = £\frac{C_1\lambda}{\mu - \lambda}$ /h

cost of service facility $C_s = £\mu C_2$ /h

total system cost $C_T = C_D + C_s$

$$= \frac{C_1\lambda}{\mu - \lambda} + \mu C_2$$

Condition for minimum cost is $dC_T/d\mu = 0$. Differentiating C_T with respect to μ

$$\frac{dC_T}{d\mu} = -\frac{C_1\lambda}{(\mu - \lambda)^2} + C_2$$

$$= 0 \text{ for minimum}$$

thus $(\mu - \lambda)^2 = \frac{C_1\lambda}{C_2}$

$$\mu = \lambda + \sqrt{\frac{C_1\lambda}{C_2}}$$

4 M/M/1/N SYSTEMS

4.1 INTRODUCTION

In this chapter the M/M/1/N system is covered - the single-channel queueing system given in chapter 3 but with a constraint that the maximum number in the system cannot exceed N. Since the arrival rate has to be random and constant, this special case is generated when the number of potential customers is infinite, but they only join the system when $n < N$; when $n = N$, the customers arriving for service go elsewhere, that is, a non-captive system.

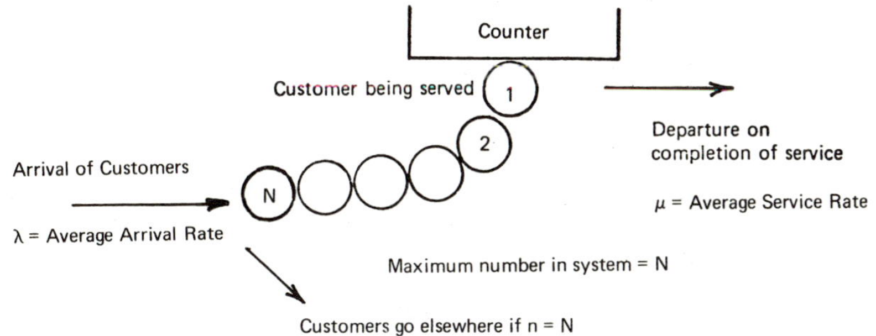

Figure 4.1 M/M/1/N System

4.2 RÉSUMÉ OF THEORY AND FORMULAE

The basic formulae for this system are given below.

M/M/1/N System

Arrival Average rate λ Poisson rate distribution

Service Average rate μ Poisson service rate distribution, neg. exp. service time distribution

Number of servers one

System size limitation max. size N

Queue discipline first in, first out

Probability of no customers in the system $P_0 = \dfrac{1 - \rho}{1 - \rho^{N+1}}$

Probability of n customers in the system $P_n = \frac{\rho^n(1 - \rho)}{1 - \rho^{N+1}}$ $(0 \leqslant n \leqslant N)$

Proportion of customers not served (non-captive system) $= P_N = \frac{\rho^N(1 - \rho)}{1 - \rho^{N+1}}$

Average number in system $\bar{n} = \rho\left[\frac{1 - (N + 1)\rho^N + N\rho^{N+1}}{(1 - \rho)(1 - \rho^{N+1})}\right]$

Average number in queue $\bar{q} = \rho^2\left[\frac{1 - N\rho^{N-1} + (N - 1)\rho^N}{(1 - \rho)(1 - \rho^{N+1})}\right]$

No formulae can be developed for average waiting time although the general formula

average waiting time $\bar{w} = \frac{\bar{n}}{\mu}$

can be used to obtain the average customer waiting time.

4.2.1 Optimisation of System under Conditions of Non-captive Customers, No Constraint on Total Potential Customers

G = Average profit per service
E = Average cost per service

Note: This condition assumes that the cost of providing the service is linear, thus doubling the service rate doubles the total cost of providing the service.

In addition, it is assumed that the service rate μ can be varied. Thus, if all customers were served

gross profit $= \lambda G$ per unit of time

However, only $(1 - P_N)$ of customers are served, therefore gross profit per unit of time is

$$\lambda G(1 - P_N) = \lambda G\left[1 - \frac{(1 - \rho)\rho^N}{1 - \rho^{N+1}}\right]$$

Net cost of service per unit of time $= E\mu$

Net profit per unit time is

$$Z = \lambda G\left(\frac{1 - \rho^N}{1 - \rho^{N+1}}\right) - E\mu = \lambda\mu G\left(\frac{\mu^N - \lambda^N}{\mu^{N+1} - \lambda^{N+1}}\right) - E\mu$$

If λ, G and E are fixed, the problem is to vary μ to maximise Z. For maximum Z, $dz/d\mu = 0$, which gives

$$\rho^{N+1}\left[\frac{N - (N + 1)\rho + \rho^{N+1}}{(1 - \rho^{N+1})^2}\right] = \frac{E}{G}$$

For given E/G and N, table 3 gives the optimum value of ρ, from which the optimum value of μ can be found.

An example of use of this theory is now given.

Example

In this example, the practical difficulty of measuring the arrival rate λ will be demonstrated. Clearly, only the customers who enter the system can be measured.

In a large department store, at the jewellery counter, there is only one assistant. Activity sampling studies of the service gave the following data on the number in the system.

No. in System	*No. of Readings*
0	138
1	286
2	576

Work study measurements of the service time gave an average of 10.5 minutes and the distribution was negative exponential.

Calculate ρ for the system and show the proportion of prospective customers lost. Calculate the expected average waiting time.

If the average profit per sale is £5 and the average cost per service is £2 calculate the optimum service rate and the increase in net profit.

Solution

This problem illustrates the more practical case where there is no simple straightforward method of measuring the average arrival rate λ, since all that can be normally measured is the input to the system. Therefore use is made of activity sampling studies and λ is estimated from the number in the system data.

Estimate of the arrival rate λ is as follows.

No. in System (n)	*No. of Readings*		*Probability (P_n)*	*Log P_n*
0		138	0.138	$\bar{1}.1399$
1		286	0.286	$\bar{1}.4564$
2		576	0.576	$\bar{1}.7604$
	Total	1000	1.000	

Since $P_n = P_0\rho^n$, then ρ can be calculated in two ways. The first where $\rho^{n-r} = (P_n/P_r)$ gives three estimates for ρ

$$\rho = \frac{286}{138} = 2.07$$

or $$\frac{576}{286} = 2.02$$

or $$\sqrt{\frac{576}{138}} = 2.04$$

The second method is a graphical one. Since $P_n = \rho^n P_0$ then $\log P_n = n \log \rho + \log P_0$ and

$$\log \rho = \frac{(\log P_n - \log P_0)}{n}$$

If $\log P_n$ is plotted against n, the points should fall on a straight line; the slope of this line gives $\log \rho$ from which ρ can be calculated. Figure 4.2 shows this plot.

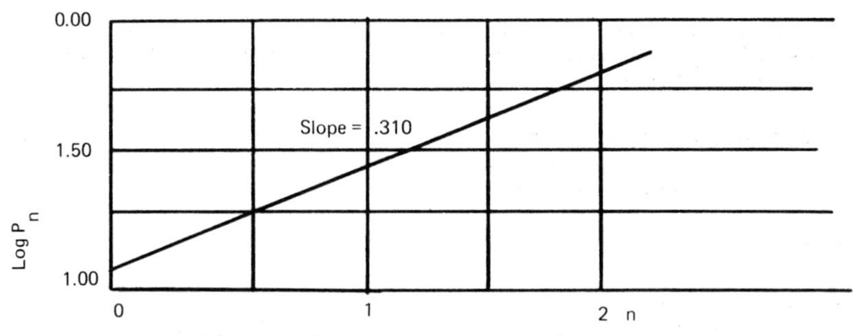

Figure 4.2 Log P_n against n

Slope = 0.310, therefore

$$\log \rho = 0.310$$

and $\rho = 2.04$

$$\text{Average service rate } \mu = \frac{60}{10.5}$$

$$= 5.71 \text{ customers per hour}$$

therefore

$$\frac{\lambda}{5.71} = 2.04$$

$$\text{Average arrival rate } \lambda = 2.04 \times 5.71$$

$$= 11.65 \text{ customers/h}$$

Maximum number in the system N = 2, therefore probability of system being idle is

$$P_0 = \frac{1 - \rho}{1 - \rho^3} = \frac{1 - 2.04}{1 - 2.04^3} = \frac{-1.04}{-7.50} = 0.14$$

$$P_1 = 0.139 \times 2.04 = 0.28$$

$$P_2 = 0.282 \times 2.04 = 0.58$$

(Cross-check: $P_0 + P_1 + P_2 = 1.00$. Thus P_2, proportion of customers who go elsewhere for service, = 0.58 or 58 per cent.)

Optimum Service Rate

Here E = £2, G = £5, N = 2. Reference to table 3 gives for optimum

$$\rho = 1.16$$

Since $\lambda = 11.62$, therefore

$$\mu = \frac{11.62}{1.16} = 10 \text{ customers/h}$$

Compare present service rate of only 5.71 customers per hour, thus the service rate should be increased.

Increase in Net Profit

Current net profit/h = Total profit/h - Cost of service/h

$$= £G\,(1 - P_N)\lambda - £E\mu$$

$$= £5 \times (1 - 0.58) \times 11.62 - £2 \times 5.71$$

$= £24.4 - 11.4$

$= £13$ per hour

With $\mu = 10$, $\rho = 1.16$, $N = 2$

$$P_0 = \frac{1 - \rho}{1 - \rho^{N+1}} = \frac{1 - 1.16}{1 - 1.16^3} = \frac{0.16}{0.56} = 0.286$$

thus $P_N = 0.286 \times 1.16^2 = 0.385$

therefore

$$\text{Optimum net profit/h} = £5\,(1 - 0.385) \times 11.62 - £2 \times 10$$

$= £35.7 - 20$

$= £15.7/\text{h}$

or an increase of £2.7/h in the net profit.

4.3 PROBLEMS

1. Show that for an M/M/1/N system the steady-state probabilities are

$$P_n = \frac{1 - \rho}{1 - \rho^{N+1}} \rho^n$$

where $\rho = \lambda/\mu$ and derive the limit values for these probabilities when $\rho \to 1$.

2. A petrol station has a single pump and space for no more than 3 cars (2 waiting, 1 being served). A car arriving when the space is filled to capacity goes elsewhere for petrol. Cars arrive according to a Poisson distribution at a mean rate of one every 8 minutes. Their service time has an exponential distribution with a mean of 4 minutes.

The proprietor has the opportunity of renting an adjacent piece of land, which would provide space for an additional car to wait. (He cannot build another pump.) The rent would be £10 per week. The expected net profit from each customer is 50p and the station is open 10 hours every day. Would it be profitable to rent the additional space?

3. Activity sampling studies at a service counter gave the following data.

No. in System	*No. of Readings*
0	1605
1	971
2	531
3	342

Average service time = 10 minutes.

Given that the arrival rate of potential customers is random, that the service time has a negative exponential distribution, also that the cost of providing one service is £1 and that the average gross profit per customer is £2, what service rate should be adopted and by how much will the net profit be increased?

What proportion of potential customers will be lost under the conditions of maximum profit?

4. Customers arrive at random at a single service station with mean arrival rate λ. However, a customer will join the system to be serviced only if there are less than N customers already in the system when he arrives, otherwise he goes elsewhere. Service times are distributed according to the negative exponential distribution with mean time $1/\mu$, and customers who join the system are served in the order in which they arrive.

Show that the expected number of customers passing through the service station in unit time is given by the expression

$$\lambda\left[\frac{1-\left(\frac{\lambda}{\mu}\right)^{N}}{1-\left(\frac{\lambda}{\mu}\right)^{N+1}}\right]$$

The above model represents a small petrol filling station for which N = 2 and μ = 30 per hour. A sum of money is available for improving the station and can be used either to buy more space to increase N to 3 or to install a faster pump to increase μ to 40 per hour. If λ is 10 per hour, which of these two alternatives will produce the greater increase in profits, assuming the average profit per customer remains constant?

If λ were 30 per hour, would the same answer be valid?

5. Consider a production process with two machines A and B. The output from the first machine is Poisson distributed, with the rate of 4 per hour. The second machine B processes each item with exponential service time at a rate of 5 per hour. There is a congestion problem when

more than 2 items are waiting to be processed on machine B.

(a) Calculate the fraction of time that the system is congested in the steady state.

(b) Suppose machine A is turned off when more than 2 items are waiting to be processed by machine B. Calculate the steady-state probability of the system being congested in this case. Compare with answer in (a).

4.4 SOLUTIONS

1. Steady-state equations are

$$-\lambda P_0 + \mu P_1 = 0 \qquad n = 0$$

$$\lambda P_{n-1} - (\lambda + \mu)P_n + \mu P_{n+1} = 0 \qquad 1 \leqslant n \leqslant N - 1$$

$$\lambda P_{N-1} - \mu P_N = 0 \qquad n = N$$

Setting $\rho = \lambda/\mu$

$$P_1 = \rho\, P_0 \qquad n = 0$$

$$P_{n+1} = (\rho + 1)P_n - \rho P_{n-1} \qquad 1 \leqslant n \leqslant N - 1$$

$$P_N = \rho P_{N-1} \qquad n = N$$

These may be solved by recursion

$$P_2 = (\rho + 1)P_1 - \rho P_0 = \rho^2 P_0$$

$$P_n = \rho^n P_0$$

From $\sum_{n=0}^{N} P_n = 1$

$$P_0 \sum_{n=0}^{N} \rho^n = P_0 \frac{1 - \rho^{N+1}}{1 - \rho} = 1$$

Therefore

$$P_n = \frac{1 - \rho}{1 - \rho^{N+1}} \rho^n$$

$$\lim_{\rho \to 1} P_n = \lim_{\rho \to 1} \rho^n \frac{(1 - \rho)}{(1 - \rho^{N+1})}$$

$$= \lim_{\rho \to 1} \rho^n \lim_{\rho \to 1} \frac{1 - \rho}{1 - \rho^{N+1}}$$

$$= 1 \lim_{\rho \to 1} \frac{1 - \rho}{1 - \rho^{N+1}}$$

The latter limit can be obtained by applying L'Hospital's rule

$$\lim_{\rho \to 1} \frac{1 - \rho}{1 - \rho^{N+1}} = \lim_{\rho \to 1} \frac{d}{d\rho} (1 - \rho) \Big/ \lim_{\rho \to 1} \frac{d}{d\rho} (1 - \rho^{N+1})$$

$$= \frac{-1}{-(N+1)} = \frac{1}{N+1}$$

2. *Present System*

Average arrival rate $\lambda = \frac{60}{8} = 7.5$ cars/h

Average service rate $\mu = \frac{60}{4} = 15$ cars/h

$$\rho = \frac{\lambda}{\mu} = \frac{7.5}{15} = 0.5$$

Maximum number in system, $N = 3$. Probability of system being empty

$$P_0 = \frac{1 - \rho}{1 - \rho^{N+1}}$$

thus $P_0 = \dfrac{1 - 0.5}{1 - (0.5)^4} = 0.533$

Probability of system in maximum state, $N = 3$, is

$$P_3 = 0.533 \times (0.5)^3$$

$$= 0.067$$

Therefore proportion of lost customers = 0.067

Proposed System

If maximum number in system increased to $N = 4$

$$P_0 = \frac{1 - 0.5}{1 - (0.5)^5} = 0.516$$

Probability of system being in maximum state is

$$P_4 = 0.516 \times (0.5)^4 = 0.032$$

thus proportion of lost customers = 0.032

therefore

Increase in cars

served per hour $= \lambda(0.067 - 0.032)$

$= 7.5 \times 0.035$

$= 0.262$ cars/h

Increase in cars

served/week $= 0.262 \times 10 \times 7$

$= 18.34$/week

Thus increase in profit per week = 50p × 18.34

= £9.17

Since rent for additional space would be £10 per week, it is not economical to increase the existing space.

3. Determine arrival rate (λ) by finding average traffic intensity (ρ). This is an M/M/1/N process. Thus

$$P_n = \frac{(1 - \rho)}{1 - \rho^{N+1}} \rho^n$$

where N = maximum number in system. Therefore

$$\frac{P_n}{P_{n-1}} = \rho$$

let x_n = number of activity sample readings and $P_n \propto x_n$. Since N = 3

$$\rho = \frac{1}{3} \sum_{n=1}^{3} \frac{x_n}{x_{n-1}}$$

(see also graphical method, p.)

$$= \frac{0.644 + 0.547 + 0.605}{3} = 0.60$$

$\rho = 0.6$, but $\mu = 6$/h, therefore

$$\lambda = \rho\mu = 3.6\text{/h}$$

The optimum values of ρ as a function of N and E/G are given in table 3. For N = 3 and E/G = 0.5, ρ = 1.21. Hence

$$\mu = \frac{\lambda}{\rho} = \frac{3.6}{1.21} = 3/\text{hour}$$

To determine the increase in profit, compute the present and the proposed profits.

$$\text{Present profit} = £2 \times 3.6 \times \left[1 - \frac{1 - 0.6}{1 - 0.6^4} \times 0.6^3\right] - 1 \times 6$$

$$= £0.46/\text{h}$$

$$\text{Proposed profit} = £2 \times 3.6 \left[1 - \frac{1 - 1.21}{1 - 1.21^4} \times 1.21^3\right] - 1 \times 3$$

$$= £1.86$$

Increase in profit = £1.86 - £0.46 = £1.40/h

Proportion of customers not served is P_N

$$P_N = \frac{(1 - \rho)\rho^N}{1 - \rho^{N+1}}$$

original customer loss P_N for ρ = 0.6, N = 3

$$P_N = \frac{(1 - 0.6) \times 0.22}{0.87} = 0.10$$

with new service rate

$$P_N = \frac{(1 - 1.21) \times 1.77}{-1.14} = 0.32 \quad (32\%)$$

Note that the solution maximises profit but, as can be seen, about one-third of the potential customers are lost.

4. Average number passing through service station = $\lambda(1 - P_N)$, but

$$P_N = \left(\frac{\lambda}{\mu}\right)^N \left[\frac{1 - \frac{\lambda}{\mu}}{1 - \left(\frac{\lambda}{\mu}\right)^{N+1}}\right]$$

therefore

$$1 - P_N = \left[\frac{1 - \left(\frac{\lambda}{\mu}\right)^N}{1 - \left(\frac{\lambda}{\mu}\right)^{N+1}}\right]$$

Therefore average number through service station per unit time $\bar{S}(n)$ is

$$\bar{S}(n) = \lambda\left[\frac{1 - \left(\frac{\lambda}{\mu}\right)^N}{1 - \left(\frac{\lambda}{\mu}\right)^{N+1}}\right]$$

(a) Consider installing a faster pump, $N = 2$, $\lambda = 10/h$, $\mu = 40/h$.

$$\bar{S}(n) = 10 \times \frac{1 - (1/4)^2}{1 - (1/4)^3}$$

$$= 10 \times \frac{15}{16} \times \frac{64}{63} = 9.5$$

that is, an average 9.5 customers will be passing through the service station per hour.

(b) Consider buying extra space, $N = 3$, $\lambda = 10/h$, $\mu = 30/h$.

$$\bar{S}(n) = 10 \times \frac{1 - (1/3)^3}{1 - (1/3)^4}$$

$$= 10 \times \frac{26}{27} \times \frac{81}{80} = 9.75$$

that is, on average 9.75 customers will be passing through the service station per hour.

Therefore, choose to purchase the extra space. Suppose $\lambda = 30/h$. For case (a), the faster pump, $N = 2$, $\lambda = 30/h$, $\mu = 40/h$.

$$\bar{S}(n) = 30 \times \frac{1 - (3/4)^2}{1 - (3/4)^3} = 22.7$$

For case (b), the extra space, $N = 3$, $\lambda = 30/h$, $\mu = 30/h$. Therefore

$$\rho = \frac{\lambda}{\mu} = 1$$

Here it is not possible to use the formulae $P_n = \rho^n P_0$

to calculate the probabilities. However, since

$$P_0 + P_1 + P_2 + P_3 = 1$$

therefore

$$P_0 \, (1 + \rho + \rho^2 + \rho^3) = 1$$

thus $P_0 = \frac{1}{4}$

and it follows that $P_1 = P_2 = P_3 = P_0 = \frac{1}{4}$

With $\lambda = 30/h$ the number of customers who go elsewhere is $30 \times P_3 = 7.5/h$. Therefore

$$\bar{S}(n) = 22.5$$

Hence it would be marginally better to install a faster pump.

5. (a) $M/M/1/\infty$ system, with the input (λ) the output from machine (A), the service rate (μ), the production rate of machine (B).

$$\lambda = 4/h$$

$$\mu = 5/h$$

$$\rho = 4/5 = 0.8$$

Prob. (system congested) = Prob. (more than 2 in queue) = Prob. (more than 3 in system)

$$\sum_{n=4}^{\infty} P_n = \rho^4 = 0.410$$

that is, the system is congested 41% of the time.

(b) Machine (A) is turned off whenever there are 4 items in the system. Therefore this is an $M/M/1/N$ system, with $N = 4$.

Prob. (system congested) = Prob. (4 in the system)

$$= \frac{(1 - \rho)}{(1 - \rho^5)} \rho^4$$

$$= \frac{0.2}{(1 - 0.8^5)} \times 0.8^4$$

$$= 0.122$$

Now the system is congested for 12.2% of the time.

5 M/M/C/ SYSTEMS

5.1 INTRODUCTION

Again the theory of these systems is to be found in most textbooks. See references at the end of book.

In the M/M/C/∞ system, customers arrive randomly for service, service time distribution is negative exponential, C servers, no constraint on queue size and customers are served in order of arrival.

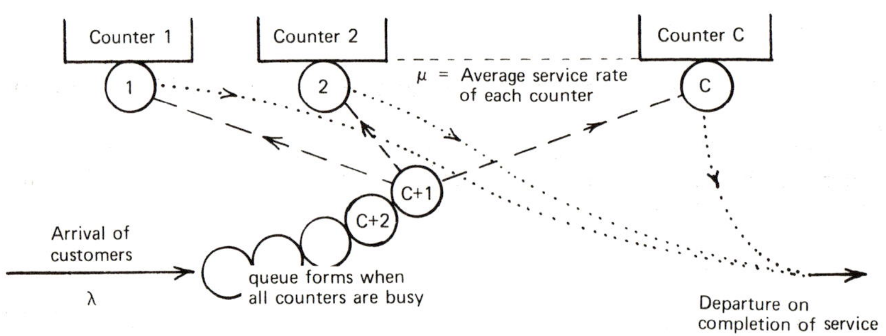

Figure 5.1 M/M/C/∞ System

5.2 RÉSUMÉ OF BASIC THEORY AND FORMULAE

The basic formulae for these systems are given below.

Probability of no customers in the system is

$$P_0 = \left[\sum_{j=0}^{c-1}\left(\frac{\lambda}{\mu}\right)^j \frac{1}{j!} + \frac{1}{c!}\left(\frac{\lambda}{\mu}\right)^c \frac{c\mu}{c\mu-\lambda}\right]^{-1}$$

Probability of n customers in the system is

$$P_n = P_0\left(\frac{\lambda}{\mu}\right)^n \frac{1}{n!} \qquad n \leqslant c$$

$$P_n = P_0\left(\frac{\lambda}{\mu}\right)^c \left(\frac{\lambda}{c\mu}\right)^{n-c} \times \frac{1}{c!} \qquad n \geqslant c$$

Average number in the queue is

$$\bar{q} = \frac{P_0 \lambda\mu(\lambda/\mu)^c}{(c-1)!(c\mu-\lambda)^2}$$

Average number in the system is

$$\bar{n} = \bar{q} + \frac{\lambda}{\mu}$$

Average waiting time is

$$\bar{w} = \frac{P_0\mu(\lambda/\mu)^c}{(c-1)!(c\mu-\lambda)^2}$$

Average time in the system is

$$\bar{d} = \bar{w} + \frac{1}{\mu}$$

Waiting time distribution is

$$w(t)dt = \frac{P_0 c\mu(\lambda/\mu)^c}{c!}e^{-(c\mu-\lambda)t}dt$$

Probability of a customer waiting longer than time t is

$$P(w>t) = \frac{P_0 c(\lambda/\mu)^c}{c!(c\mu-\lambda)}e^{-(c\mu-\lambda)t}$$

5.3 PROBLEMS

1. Arrivals at a telephone booth are considered to be Poisson, with an average time of 10 minutes between one arrival and the next. The length of a 'phone call is assumed to be distributed exponentially, with an average time of three minutes.

(a) What is the probability that a person arriving at the booth will have to wait?

(b) What is the average length of the queue that will form from time to time?

(c) The G.P.O. will install a second telephone when convinced that a customer would expect to have to wait at least 3 minutes for the 'phone. By how much must the arrival rate increase in order to justify the installation of a second 'phone booth?

(d) What is the probability that a customer will have to wait more than 10 minutes for the 'phone?

(e) If the second telephone is installed what will the average waiting time now be per customer?

2. A small ship-building company has s slipways on which it builds cabin cruisers and yachts, of a wide range of tonnages. The average building time is $1/\mu$ and may be considered to be negative exponentially distributed. Orders for ships have an average inter-arrival time of $1/\lambda$, which is also negative exponentially distributed. The orders are dealt with in order of arrival.

Write down the steady-state probability difference equations, and show that the expected number of slipways in use at any moment in time is independent of the number of slipways (provided that the system is not overloaded).

If there are 4 slipways, calculate the probability that an order has to wait for a slipway to become available, given that the average time on the slipway is one calendar month and there are an average of 12 orders per year.

3. A single server in a supermarket has a service rate of 30 customers per hour, and the service time distribution is negative exponential.

Given that an acceptable average waiting time per customer is 2 minutes, calculate at what customer random arrival rates more assistants should be put on to serving duties.

Draw a graph to show, for random arrivals, random service time, several servers, the relationship between waiting time, arrival rate and number of servers.

4. A company manufactures expensive custom-built precision instruments. Before delivery the instruments are inspected, and instruments arrive for inspection according to a Poisson distribution at a mean rate λ of one every 100 minutes. At present the company employs one inspector and the inspection is distributed negative exponentially with mean rate μ of one every 90 minutes. The average value of the instruments is £5000, and under the present system the value of work in progress awaiting future inspection is high. The management consider they can obtain 25 per cent rate of return in alternative investments, if capital tied up in this stock can be reduced.

If an additional inspector costs the company £6000 per year, would it be economical to employ a second inspector?

If a second inspector is employed what is the proportion of time they would be working on average?

5. A machine-process has an exponentially distributed service time with mean $1/\mu$. Items arrive randomly for processing at a rate λ; $\{\lambda/\mu<1\}$. At present, management think there is much congestion and two proposals have been suggested.

(a) To buy a second similar machine-process station to be operated in parallel;

(b) To replace the present machine-process station with a machine-process station of twice the service rate (but still exponentially distributed).

Given that the probability of an item having to wait at present is equal to the traffic intensity, by setting up the steady-state probability difference equations for the two situations, determine which of the two alternatives will produce the greater reduction in the steady-state average queue length.

If jobs were to arrive at the rate of 15 per hour, and the present machine-process can handle 16 per hour, what would be the expected average queue length for the present system and its expected value under each of the proposed alternatives?

6. In a self-service store the arrival process is Poisson with 9 customers arriving every 5 minutes on average. A single cashier can serve 10 customers every 5 minutes on average with time distributed negative exponentially.

The store manager wishes to reduce the probability that customers have to wait more than 1 minute. In order to do this he can either double the service rate by providing an additional server to pack the customers goods or by providing an additional cash desk. Which scheme is preferable?

7. In a machine shop the following study was carried out on a grinding machine installed for the use of operators (number of operators = 100): the average time taken to grind tools was 10 minutes and this time was negatively exponentially distributed; the average number of calls per week for grinding was 200 in a 44-hour week.

What is the total time lost by operators, in one week, in waiting in the queue for the grinding machine?

If a second machine is installed, by how much will the total waiting time be reduced?

8. A company is studying a proposed operation involving a new fleet of trucks and an unloading depot, at which it is estimated they will arrive in a random fashion at an average rate of 3 per hour. It is proposed to construct either one or two unloading bays. The average time taken to unload a truck at an unloading bay is estimated to be 15 minutes. A queue of lorries for unloading will form from time to time. The capital cost of each truck will be £8,000. The cost of constructing

the second unloading bay would be £10,000. Consider the question whether the construction of the second unloading bay would be worthwhile, given that the unloading time is negatively exponentially distributed.

5.4 SOLUTIONS

1. λ = 0.10 arrivals per minute, μ = 0.33 calls per minute therefore

$$\rho = \frac{\lambda}{\mu} = 0.3$$

(a) Probability that an arrival has to wait = probability that the telephone is in use = probability that there is at least one person in the system = 1 - probability that there is no one in the systen

$$= 1 - P_0$$

$$= 1 - (1 - \rho) = \rho = 0.3$$

(b) Average length of queue

$$\bar{q} = \frac{\rho^2}{1-\rho} = \frac{0.3^2}{1-0.3} = 0.13 \text{ persons}$$

(c) Average or expected waiting time

$$\bar{w} = \frac{\lambda}{\mu(\mu-\lambda)}$$

$$= \frac{\lambda}{0.33(0.33-\lambda)} \qquad \text{with } \mu = 0.33$$

Equating $3 = \lambda/\{0.33(0.33-\lambda)\}$ gives λ = 1/6 arrivals per minute. Thus a second booth is only justified, according to the stated criterion, if the arrival rate per hour is increased from 6 to 10 or more.

(d) $P(w > t)$ = probability of waiting more than time t in queue

$$= \rho e^{-t(\mu-\lambda)}$$

$$= 0.3e^{-10(0.33-0.1)} = 0.3e^{-2.3} = 0.03$$

(e) This is an M/M/2/∞ System. From section 5.2, replacing c by 2

$$\bar{w} = \frac{P_0\mu(\lambda/\mu)^2}{1!(2\mu-\lambda)^2}$$

and

$$P_0 = \left[1 + \frac{\lambda}{\mu} + \frac{1}{2!}\left(\frac{\lambda}{\mu}\right)^2 \frac{2\mu}{2\mu-\lambda}\right]^{-1}$$

With the substitutes $\lambda = 0.1$, $\mu = 0.33$ and $\lambda/\mu = 0.3$

$$P_0 = \left[1 + 0.3 + \frac{(0.3)^2}{2} \frac{0.66}{0.66-0.1}\right]^{-1}$$

$$= (1.35)^{-1} = 0.74$$

Average waiting time is

$$\bar{w} = 0.74 \frac{0.33(0.3)^2}{(0.66-0.1)^2} = \frac{0.022}{0.313} = 0.07 \text{ min} = 4.2 \text{ sec}$$

Or Average waiting time, with two booths, is 4.2 sec.

2\. This is an $M/M/C/\infty$ system with $C = s$. The steady-state equations are

$$- \lambda P_0 + \mu P_1 = 0$$

$$- (\lambda+\mu)P_1 + \lambda P_0 + 2\mu P_2 = 0$$

$$- (\lambda+2\mu)P_2 + \lambda P_1 + 3\mu P_3 = 0$$

$$\vdots \qquad \vdots \qquad \vdots$$

$$- (\lambda+n\mu)P_n + \lambda P_{n-1} + (n+1)\mu P_{n+1} = 0 \qquad n < s$$

$$- (\lambda+s\mu)P_s + \lambda P_{s-1} + s\mu P_{s+1} = 0 \qquad n = s$$

$$- (\lambda+s\mu)P_n + P_{n-1} + s\mu P_{n+1} = 0 \qquad n > s$$

Solving the equations recursively gives

$$P_n = \frac{(s\rho)^n}{n!} P_0 \qquad n \leqslant s$$

$$P_n = \frac{s^s}{s!}\rho^n P_0 \qquad n \geqslant s$$

where $\rho = \lambda/(s\mu)$.

Now, expected number of slipways in use = expected number of customers in service = expected number in system - expected number in queue, that is

$$\bar{n} - \bar{q} = \sum_1^\infty nP_n - \sum_{s+1}^\infty (n-s)P_n$$

$$= \sum_{1}^{s} nP_n + \sum_{s+1}^{\infty} nP_n - \sum_{s+1}^{\infty} nP_n + s\sum_{s+1}^{\infty} P_n$$

$$= \sum_{1}^{s} nP_n + s\sum_{s+1}^{\infty} P_n$$

$$= \sum_{1}^{s} \frac{(s\rho)^n}{(n-1)!}P_0 + s\sum_{s+1}^{\infty} \frac{s^s}{s!}\rho^n P_0$$

$$E(s) = s\rho P_0\left[\left(1 + s\rho + \frac{(s\rho)^2}{2!} + \ldots + \frac{(s\rho)^{s-1}}{s-1!}\right) + \left(\frac{s^s}{s!}\rho^s + \frac{s^s}{s!}\rho^{s+1} + \ldots\right)\right]$$

therefore

$$E(s) = s\rho P_0 \frac{1}{P_0} = s\rho$$

since $\rho = \lambda/(s\mu)$, then $E(s) = \lambda/\mu$, which is independent of the number of slipways. With $s = 4$

probability that an order has to wait $= P_4 + P_5 + P_6 + \ldots$

$$P_n = \frac{4^4}{4!}\left(\frac{\lambda}{4\mu}\right)^n P_0 \qquad n \geq 4$$

$$= \left(\frac{1}{4}\right)^{n-4} \frac{1}{4!}P_0 \qquad \text{since } \frac{\lambda}{\mu} = 1$$

$$P_0 = \left[\sum_{j=0}^{3}\left(\frac{\lambda}{\mu}\right)^j \frac{1}{j!} + \frac{1}{4}\left(\frac{\lambda}{\mu}\right)^4 \frac{4}{4\mu-\lambda}\right]^{-1}$$

$$= \left[\sum_{j=0}^{3}\frac{1}{j!} + \frac{1}{4!}\,\frac{4}{(4-1)}\right]^{-1}$$

$$= \left(\frac{16}{6} + \frac{1}{18}\right)^{-1} = \left(\frac{49}{18}\right)^{-1}$$

and

$$P_0 = \frac{18}{49}$$

therefore, probability of wait $= \sum_{n=4}^{\infty} P_n$

$$= P_0 \frac{1}{4!}\sum_{n=4}^{\infty}\left(\frac{1}{4}\right)^{n-4}$$

$$= P_0 \frac{1}{4!} \sum_{n=0}^{\infty} \left(\frac{1}{4}\right)^n$$

$$= P_0 \frac{1}{4!} \left(\frac{1}{1-1/4}\right)$$

$$= \frac{4}{3\times 4!} P_0 = \frac{P_0}{18} = \frac{1}{49}$$

3. In order to obtain the arrival rate giving an average waiting time $\bar{w}$ of 2 minutes, we must compute the waiting times and derive the result by approximation. The simplest method for $c > 1$ (that is, more than one server) is to determine values of $1/\bar{w}$ the reciprocal of average waiting time.

Let $\varepsilon = \lambda/\mu$. For an M/M/C/$\infty$ process

$$\bar{w} = P_0 \frac{\varepsilon^c}{(c-1)!\mu(c-\varepsilon)^2}$$

and $P_0 = \left[\sum_{j=0}^{c-1} \frac{\varepsilon^j}{j!} + \frac{\varepsilon^c}{(c-1)!(c-\varepsilon)}\right]^{-1}$

Therefore

$$\frac{1}{\bar{w}} = \mu(c-\varepsilon)\left[1 + (c-\varepsilon)\sum_{j=0}^{c-1} \frac{(c-1)!}{j!} \times \frac{1}{\varepsilon^{c-j}}\right]$$

$$= \mu(c-\varepsilon)\left[1 + \left(\frac{1}{\varepsilon} + \frac{(c-1)}{\varepsilon^2} + \frac{(c-1)(c-2)}{\varepsilon^3} + \ldots + \frac{(c-1)!}{\varepsilon^c}\right)(c-\varepsilon)\right] \qquad 5.1$$

For $c > 1$, this expression reduces to

$$\frac{1}{\bar{w}} = \mu(c-\varepsilon) \frac{c}{\varepsilon} + \left(\frac{c}{\varepsilon} - 1\right) \times \sum_{j=0}^{c-2} \prod_{k=j+1}^{c-1} \frac{k}{\varepsilon} \qquad 5.2$$

and for $c = 1$

$$\frac{1}{\bar{w}} = \frac{\mu(1-\varepsilon)}{\varepsilon} \qquad \text{from } 5.1$$

The following results were calculated on the Olivetti (101) and give the graph shown in figure 5.2.

No. of Servers	*1*	*2*	*3*	*4*	*5*
Upper limit of λ	15	42.4	71.1	100.2	129.5
Increase in λ before change	27.4	28.7	29.1	29.3	

4.

(a) *Present System*

Rate of arrival $= \frac{1}{100}$ /min

Rate of inspection $= \frac{1}{90}$ /min

thus $\rho = \frac{\lambda}{\mu} = 0.9$

Average number of instruments in the system

$$\bar{n} = \frac{\rho}{1-\rho} = 9$$

(b) *Proposed System (Two Inspectors)*

From section 5.2, average number of instruments in inspection system is

$$\bar{n} = \bar{q} + \frac{\lambda}{\mu}$$

where $\bar{q} = \dfrac{P_0 \lambda\mu(\lambda/\mu)^2}{(2\mu-\lambda)^2 1!}$ with n = 2

and $P_0 = \left[1 + \frac{\lambda}{\mu} + \frac{1}{2!}\left(\frac{\lambda}{\mu}\right)^2 \frac{2\mu}{2\mu-\lambda}\right]^{-1}$

that is

$$P_0 = \left[1 + (0.9) + \frac{1}{2}(0.9)^2 \frac{\frac{2}{90}}{\frac{2}{90} - \frac{1}{100}}\right]^{-1}$$

$$= \left[1 + (0.9) + \frac{1}{2}(0.81) \frac{2}{2-0.9}\right]^{-1}$$

$$= (2.64)^{-1} = 0.379$$

To calculate λ from μ during programme runs:

$$\left\{ \begin{array}{ll} \lambda/\mu(\epsilon)\ 2.37 & \downarrow X \\ \mu \rightarrow 30 & \\ \lambda{-}71.1000 & A \end{array} \right.$$

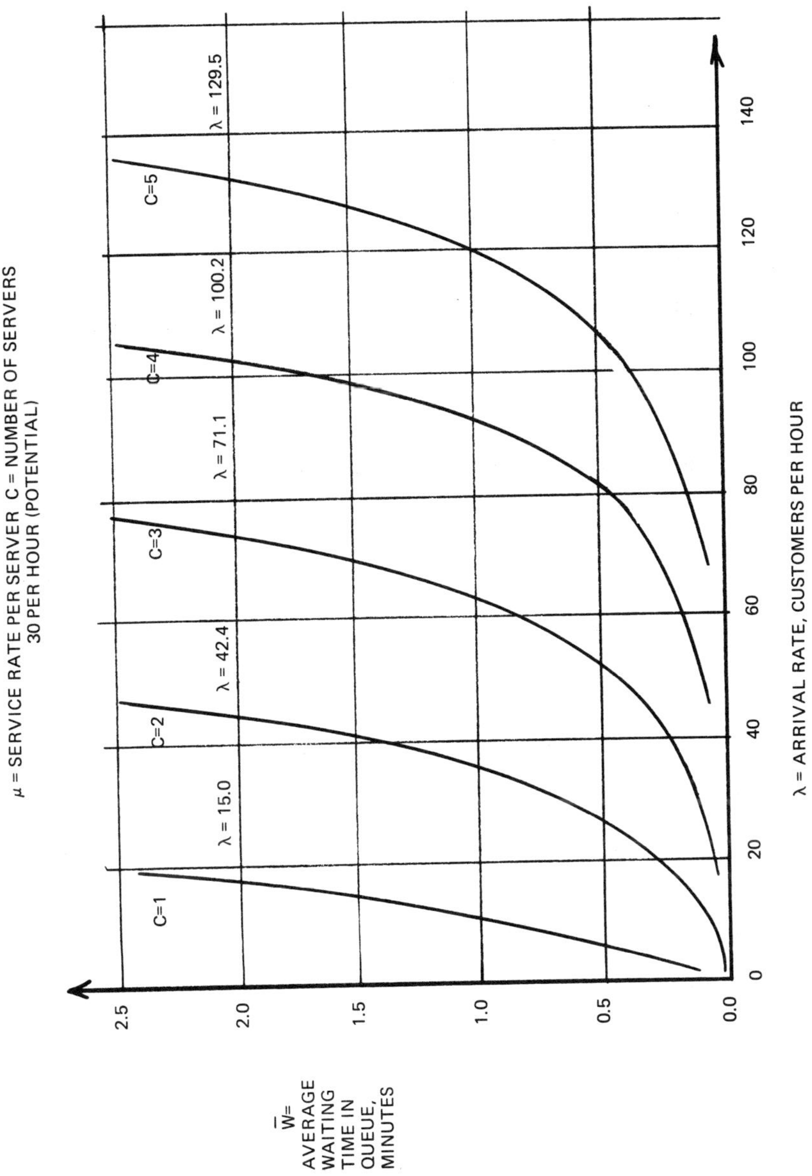

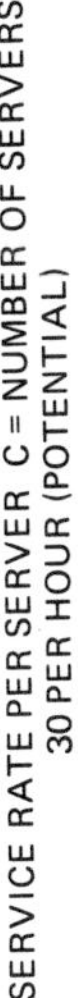

Figure 5.2

$$\bar{q} = \frac{0.379 \times \frac{1}{100} \times \frac{1}{90} (0.9)^2}{1!\left(\frac{2}{90} - \frac{1}{100}\right)^2}$$

$$= 0.379 \frac{(0.9)^2 \times 90 \times 100}{(200-90)^2}$$

$$= 0.379 \frac{7290}{12100}$$

$$= 0.379 \times 0.602$$

$$= 0.23$$

So that the average number in the system is

$$\bar{n} = 0.23 + 0.9$$

$$= 1.13$$

Reduction in number of units in work in progress

$$= 9 - 1.13 = 7.87$$

Average value of this reduction = 7.87 × £5000 = £39350

Annual reduction in stockholding costs = £39350 × $\frac{25}{100}$

= £9837.5

Net annual saving by employing second inspector

= £9837.5 - £6000 = £3837.5

For a system with c servers the average utilisation per server is

$$\rho = \frac{\lambda}{c\mu} = \frac{0.90}{2} = 0.45$$

Therefore, on average, the inspectors will be working 45 per cent of the time.

5.

(a) *Second Machine*

Installing an additional machine.

The steady-state difference equations are

$$-\lambda P_0 + \mu P_1 = 0$$

$$\lambda P_0 - (\lambda + \mu)P_1 + 2\mu P_2 = 0$$

$$\lambda P_1 - (\lambda + 2\mu)P_2 = 2\mu P_3 = 0$$

$$P_2 - (\lambda + 2\mu)P_3 + 2\mu P_4 = 0$$

etc.

Solving this equations recursively gives

$$P_1 = \frac{\lambda}{\mu} P_0$$

$$P_2 = \frac{1}{2}\left(\frac{\lambda}{\mu}\right)^2 P_0$$

or $$P_n = \frac{1}{2^{(n-1)}} \left(\frac{\lambda}{\mu}\right)^n P_0$$

Let $\frac{\lambda}{\mu} = \varepsilon$;

To obtain P_0

$$P_0 + P_1 + P_2 + \ldots = 1$$

thus $$P_0\left(1 + \varepsilon + \frac{1}{2}\varepsilon^2 + \frac{1}{2^2}\varepsilon^3 + \ldots\right) = 1$$

$$P_0 \; 2\left[\frac{1}{2} + \left(\frac{\varepsilon}{2}\right) + \left(\frac{\varepsilon}{2}\right)^2 + \left(\frac{\varepsilon}{2}\right)^3 + \ldots\right] = 1$$

$$P_0\left\{-1 + 2\left[1 + \left(\frac{\varepsilon}{2}\right) + \left(\frac{\varepsilon}{2}\right)^2 + \left(\frac{\varepsilon}{2}\right)^3 + \ldots\right]\right\} = 1$$

$$P_0\left(\frac{2}{1-\varepsilon/2} - 1\right) = 1$$

thus $$P_0 = \frac{2-\varepsilon}{2+\varepsilon}$$

Average number in the queue is

$$\overline{q} = P_3 + 2P_4 + 3P_5 + \ldots$$

$$= P_0\left(\frac{1}{2^2}\varepsilon^3 + 2\,\frac{1}{2^3}\varepsilon^4 + 3\frac{1}{2^4}\varepsilon^5 + \ldots\right)$$

$$= P_0 \frac{\varepsilon^3}{4} \left[1 + 2\left(\frac{\varepsilon}{2}\right) + 3\left(\frac{\varepsilon}{2}\right)^2 + 4\left(\frac{\varepsilon}{2}\right)^3 + \ldots\right]$$

$$= P_0 \frac{\varepsilon^3}{4} \left(1 - \frac{\varepsilon}{2}\right)^{-2}$$

$$= \frac{2-\varepsilon}{2+\varepsilon}\left(\frac{\varepsilon}{4}\right)^3 \frac{4}{(2-\varepsilon)^2} = \frac{\varepsilon^3}{(2+\varepsilon)(2-\varepsilon)} \qquad 5.3$$

(b) *One Machine Twice as Fast*

$$\bar{q} = \frac{\rho^2}{1-\rho} \quad \text{where } \rho = \left(\frac{\lambda}{2\mu}\right)$$

$$= \frac{\left(\frac{\lambda}{2\mu}\right)}{1 - \frac{\lambda}{2\mu}} = \frac{\varepsilon^2}{2(2-\varepsilon)} \qquad 5.4$$

Subtract equation in 5.3 from equation 5.4

$$\frac{\varepsilon^2(2+\varepsilon) - 2\varepsilon^3}{2(2-\varepsilon)(2+\varepsilon)} = \frac{\varepsilon^2(2-\varepsilon)}{2(2-\varepsilon)(2+\varepsilon)} = \frac{\varepsilon^2}{2(2+\varepsilon)} \qquad 5.5$$

Equation 5.5 is positive, hence two machines in parallel give a shorter queue. Original $\lambda = 15$, $\mu = 16$, therefore $\varepsilon = 15/16$.

One Old Machine

$$\bar{q} = \frac{(15/16)^2}{(1-15/16)} = \frac{15^2}{16} = \frac{225}{16} = 14$$

One Faster Machine

$$\bar{q} = \frac{(15/16)^2}{2(2-15/16)} = \frac{225}{2\times17\times16} = 0.413$$

Two Machines

$$\bar{q} = \frac{(15/16)^3}{(2+15/16)(2-15/16)} = \frac{15^3}{16\times47\times17}$$

$$= \frac{3375}{12784} = 0.264$$

6. *Double Service Rate*

$\lambda = 1.8$, $\mu = 4$, $\rho = 0.45$.

$$P(w>t) = \rho e^{-(\mu-\lambda)t}$$

$$\text{Thus } P(w>1) = 0.45\ e^{-2.2}$$

$$= 0.45 \times 0.1108$$

$$= 0.05$$

Two Cashiers

$c = 2;\ \lambda = 1.8;\ \mu = 2;\ \varepsilon = \lambda/\mu = 0.9;\ \rho = \varepsilon/c = 0.45$

$$P(w > t) = P_0 \frac{(c\rho)^c}{c!(1-\rho)} e^{-c\mu(1-\rho)t}$$

$$= P_0 \frac{c\varepsilon^c}{c!(c-\varepsilon)} e^{-(c\mu-\lambda)t}$$

$$P(w > 1) = \frac{2-\varepsilon}{2+\varepsilon} \times \frac{2\varepsilon^2}{2(2-\varepsilon)}\ e^{-(2\mu-\lambda)}$$

$$= \frac{0.9^2}{2.9}\ e^{-2.2} = \frac{0.81}{2.9} \times 0.1108$$

$$= 0.031$$

Therefore 1 customer in 32 will have to wait longer than 1 minute. Thus, if it is the store manager's objective to minimise the probability of waiting more than 1 minute for service he should install 2 cash desks.

7. *Single Grinding Machine*

Requirement for grinding machine $\lambda = 200$/week, rate of grinding $\mu = 6$/h $= 264$/week, therefore

$$\rho = \frac{\lambda}{\mu}$$
$$= \frac{200}{264} = 0.76$$

$$\text{Average waiting time } \bar{w} = \frac{\rho}{\mu(1-\rho)} \text{ weeks}$$

$$= 0.0118 \text{ weeks}$$

Thus $200 \times 0.118 = 2.36$ man-weeks are lost due to waiting per week.

Two Grinding Machines

This is an $M/M/2/\infty$ system.

From section 5.2

$$\bar{w} = P_0 \frac{\mu(\lambda/\mu)^2}{(2\mu-\lambda)^2}$$

and

$$P_0 = \left[\sum_{j=0}^{1} \left(\frac{\lambda}{\mu}\right)^j \frac{1}{j!} + \frac{1}{2!}\left(\frac{\lambda}{\mu}\right)^2 \frac{2\mu}{2\mu-\lambda}\right]^{-1}$$

Substituting $\lambda = 200$, $\mu = 264$

$$\lambda/\mu = (0.76)$$

$$P_0 = \left[1 + 0.76 + \frac{1}{2!}(0.76)^2 \frac{528}{328}\right]^{-1}$$

$$= \left[1.76 + \frac{1}{2!} \times 0.57 \times 1.61\right]^{-1}$$

$$= \left[1.76 + 0.46\right]^{-1} = (2.22)^{-1}$$

therefore

$$P_0 = 0.45$$

therefore

$$\bar{w} = 0.45 \times \frac{264 \times (0.76)^2}{(2 \times 264 - 200)^2} = 0.0006 \text{ weeks}$$

Man-weeks lost in waiting: $200 \times \bar{w} = 0.12$ weeks

Reduction in time lost

$$2.36 - 0.12 = 2.24 \text{ weeks}$$

8. *One Unloading Bay*

Average arrival rate $\lambda = 3/\text{h}$, average unloading rate $\mu = 4/\text{h}$.

$$\rho = \frac{\lambda}{\mu} = 0.75$$

Average number of trucks in the system is

$$\bar{n} = \frac{\rho}{1-\rho} = \frac{0.75}{1-0.75} = 3 \text{ trucks}$$

Average waiting time per truck is

$$\bar{w} = \frac{\rho}{(1-\rho)\mu} = 3 \times \frac{1}{3} = 1 \text{ h}$$

Two Unloading Bays

From Section 5.2

$$P_0 = \left[\sum_{j=0}^{1} \left(\frac{\lambda}{\mu}\right)^j \frac{1}{j!} + \frac{1}{2!} \left(\frac{\lambda}{\mu}\right)^2 \frac{2\mu}{2\mu-\lambda} \right]^{-1}$$

$$= \left[1 + 0.75 + \frac{1}{2!} (0.75)^2 \frac{2\times 4}{2\times 4-3} \right]^{-1}$$

$$= 0.45$$

Average number in the queue is

$$\bar{q} = P_0 \frac{\lambda\mu(\lambda/\mu)^C}{(C-1)!(C\mu-\lambda)^2}$$

$$= \frac{0.45\times 3\times 4\times (0.75)^2}{1\times (2\times 3-4)^2} = 0.12$$

Thus

Average number of trucks in the system is

$$\bar{n} = 0.12 + 0.75 = 0.87$$

Average waiting time is

$$\bar{w} = \frac{\bar{q}}{\lambda} = \frac{0.12}{3} = 0.04 \text{ h}$$

The construction of an additional loading bay reduces the number of trucks in the system by (3-0.87) or 2.13 trucks. Thus for an investment of £10,000, two trucks (2.13 exactly) are released for haulage duties. Since the trucks are worth £8,000 each, it seems a worthwhile investment (providing the two extra trucks do not lie idle).

6 SYSTEMS WITH ARRIVAL RATE AND/OR SERVICE RATE DEPENDENT ON THE NUMBER IN THE SYSTEM (M_n/M_n/-/- SYSTEMS)

6.1 INTRODUCTION

There are a large number of practical problems that are systems with either the arrival rate and/or service rate dependent on the number in the system. For example, consider a maintenance gang responsible for five large machines each of which breaks down randomly, at the average rate of once per week. Clearly the total arrival rate depends on the number already broken down (number in the system); again problems of telephone switchboards, car hire, incentives to service personnel, some stock control models are similar typical systems.

6.2 RÉSUMÉ OF BASIC THEORY AND FORMULAE

These models have basic random arrival and negative exponential distribution of service time, with the additional condition that the average arrival rate and/or the average service rate are a function of the number in the system. Thus, let λ_n = arrival rate with n in the system and μ_n = service rate with n in the system. The systems are otherwise general, that is, they can be finite or infinite systems and either single- or multi-channel systems. These systems are represented by

$$M_n/M_n/-/-$$

the dashes indicating that there is no constraint on these conditions.

These problems are solved using the following formula: probability of n customers in the system is

$$P_n = \left(\frac{\lambda_0\lambda_1 \cdots \lambda_{n-1}}{\mu_1\mu_2 \cdots \mu_n}\right) P_0$$

The average waiting time can be obtained by using the general formulae

$$\bar{w} = \frac{\bar{q}}{\bar{\lambda}}$$

$$\bar{w} = \frac{\bar{n}}{\bar{\mu}}$$

The wide area of practical application of this model is clearly illustrated in the problems for solution in section 6.4, and the special applications in section 6.3.

6.3 SPECIAL APPLICATIONS OF THEORY

The use of this theory will now be demonstrated for four special cases.

1. *Derivation of Basic Formulae for M/M/1/N Systems*

Here $\lambda_n = \lambda \qquad 0 \leqslant n < N$

$\qquad = 0 \qquad n \geqslant N$

also $\mu_n = \mu \qquad$ all n

$$P_n = \frac{\lambda_0 \cdots \lambda_{n-1}}{\mu_1 \cdots \mu_n} P_0$$

therefore

$$P_1 = \frac{\lambda}{\mu} P_0$$

$$P_2 = \left(\frac{\lambda}{\mu}\right)^2 P_0$$

$$P_N = \left(\frac{\lambda}{\mu}\right)^N P_0$$

thus $1 = \sum_{i=0}^{N} P_i$

therefore

$$1 = P_0\left[1 + \frac{\lambda}{\mu} + \left(\frac{\lambda}{\mu}\right)^2 + \dots + \left(\frac{\lambda}{\mu}\right)^N\right]$$

$$= P_0\left[\frac{(1-\rho^{N+1})}{1-\rho}\right]$$

thus $P_0 = \dfrac{1-\rho}{1-\rho^{N+1}}$

and $P_n = \dfrac{1-\rho}{1-\rho^{N+1}} \rho^n$

(see chapter 4.)

2. *Derivation of Basic Formulae for M/M/C/∞ Systems*

If number in system $n \leq c$ then there are no customers waiting. However, if $n > c$, then c customers are being served and $(n-c)$ are queueing for service. Thus

$$\lambda_n = \lambda \qquad \text{all } n$$

$$\mu_n = n\mu \qquad 0 \leq n < c$$

$$= c\mu \qquad n \geq c$$

therefore

$$P_n = \left(\frac{\lambda_0 \cdots \lambda_{n-1}}{\mu_1 \cdots \mu_n}\right) P_0$$

Thus $P_1 = \frac{\lambda}{\mu} P_0$

$$P_2 = \left(\frac{\lambda}{\mu}\right) \frac{\lambda}{2\mu} P_0$$

$$P_n = \left(\frac{\lambda}{\mu}\right)^n \frac{1}{n!} P_0 \qquad n < c$$

and $P_n = \left(\frac{\lambda}{\mu}\right)^n \frac{1}{c!\ c^{n-c}} P_0 \qquad n \geq c$

(see chapter 5.)

3. *Machine Interference*

This general problem is stated as follows: A single operator is minding m machines; this system is shown diagramatically in figure 6.1.

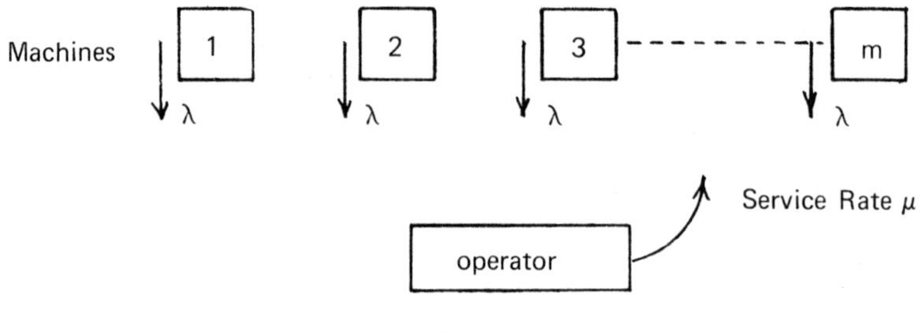

Figure 6.1

The system can be regarded as a single channel (one operator) with n customers. Thus

$$\mu_n = \mu \qquad \text{all } n$$

but $\lambda_n = (m-n)\lambda \qquad 0 \leqslant n < m$

$$= 0 \qquad n = m$$

$$P_n = \frac{\lambda_0 \ldots \lambda_{n-1}}{\mu_1 \ldots \mu_n} P_0$$

$$P_1 = \frac{m\lambda}{\mu} P_0$$

$$P_2 = m(m-1)\left(\frac{\lambda}{\mu}\right)^2 P_0$$

etc.

Thus $1 = \sum_{i=0}^{m} P_i$

$$= P_0 \left[1 + m\left(\frac{\lambda}{\mu}\right) + m(m-1)\left(\frac{\lambda}{\mu}\right)^2 \ldots m!\left(\frac{\lambda}{\mu}\right)^m \right]$$

Solving for P_0 enables a solution to be obtained for all P_n. The following can then be evaluated

$$\text{Operator utilisation} = 1 - P_0$$

$$\text{Machine utilisation} = \frac{1-P_0}{m\rho}$$

where $\rho = \lambda/\mu$. Average time a machine is in the system (broken down) is

$$\bar{d} = \frac{m}{\mu(1-P_0)} - \frac{1}{\lambda}$$

6.4 PROBLEMS

1. A skilled setter controls 5 transfer machines. Calls for resetting occur randomly on each machine with an average per machine of 1 per hour of actual running time (a machine is stopped as soon as resetting is required). The distribution of setting time is negative exponential with an average of 6 minutes.

(a) What is the probability of no machines being set up at any given time?

(b) What is the average delay for any machine in waiting for the start of resetting?

2. With reference to a process characterised by random arrivals and random departures, and using λ_n for the parameter associated with an arrival when the system is in state n and μ_n for that associated with a departure, derive the expression for the equilibrium probability that the system is in state n.

Explain what type of system corresponds to the following expression for λ_n, μ_n.

(a) $\lambda_n = \lambda \quad \mu_n = n\mu$

(b) $\lambda_n = \lambda \quad \mu_n = n\mu \ (n \leqslant c), \ \mu_n = c\mu \quad (n > c)$

(c) $\lambda_n = \dfrac{\lambda}{n+1} \quad \mu_n = \mu$

(d) $\lambda_n = \lambda (n \leqslant k-1), \ \lambda_n = 0 \ (n \geqslant k) \qquad \mu_n = \mu$

(e) $\lambda_n = \lambda (n \leqslant k-1), \ \lambda_n = 0 \ (n \geqslant k) \qquad \mu_n = n\mu (n \leqslant k)$

Prove that in case (e)

$$P_n = \frac{(1/n!)\rho^k}{1 + \rho + \frac{1}{2!}\rho^2 + \ldots + \frac{1}{k!}\rho^k}$$

where P_n = probability that the system is in state n and $\rho = \lambda/\mu$ = traffic offering.

3. A company has a spares-replacement policy, which is to order a spare when one is used from stock. The usage rate is random with mean μ per unit time and the replacement time has an exponential distribution with mean $1/\lambda$ units of time. When there is no stock, required spares are purchased immediately from a local supplier at a premium.

If the maximum stock is N, show that the system can be represented by a queueing process with random arrivals and exponential service time distribution in which the mean arrival is

$$\lambda_n = (N-n)\lambda \qquad 0 \leqslant n \leqslant N$$

$$= 0 \qquad n > N$$

and the service rate is

$$\mu_n = \mu \qquad n \geqslant 1$$

where n is the number of spares in stock.

Hence show that the steady-state probability of having n spares in stock is

$$P_n = \frac{N!}{(N-n)!}\left(\frac{\lambda}{\mu}\right)^n\left[1 + \frac{N!}{(n-1)!}\,\frac{\lambda}{\mu} + \frac{N!}{(n-2)!}\left(\frac{\lambda}{\mu}\right)^2 + \ldots\right.$$

$$\left.\ldots\quad N!\left(\frac{\lambda}{\mu}\right)^N\right]^{-1} \qquad 0 \leqslant n \leqslant N$$

$$P_n = 0 \qquad n > N$$

If (a) the mean usage rate is 2 spares per week, and the mean replacement time is 0.5 per week, (b) the cost of a spare is normally £100, but costs £115 if bought specially when no stocks are available, and (c) the cost of stock holding is 0.5% per week, what is the maximum stock level that minimises average total weekly cost?

4. A single-server queueing process is such that customers arrive at a mean rate λ per unit time, but only join the system at a rate λ_n where n is the number in the system.

Given that the service time distribution is negative exponential with mean $1/\mu$, queue discipline is first-come-first-served and that $\lambda_n = \lambda/(n+1)$, derive an expression for the expected number of customers in the system in the steady state, and show that the expected queue length is

$$(\lambda/\mu) - 1 + e^{-\lambda/\mu}$$

5. A shipping company has a single unloading berth with ships arriving in a Poisson fashion at an average rate of three per day. The unloading time distribution for a ship with n unloading crews is found to be exponential with average unloading time 1/2n days. The company has a large labour supply without regular working hours, and to avoid long waiting lines, the company has a policy of using as many unloading crews on a ship as there are ships waiting in line or being unloaded.

(a) Under these conditions, what will be the average number of unloading crews working at any time?

(b) What is the probability that more than 4 crews will be needed?

6.5 SOLUTIONS

1. The system is said to be empty when all the machines are running, that is, no machines are being set up, and

the system is full when one machine is being set up and four machines are waiting to be set up.

The arrival rate is dependent on queue size, therefore let $\lambda_n = (5-n)\lambda$; n = 0, 1, 2, 3, 4, 5. The steady-state equations are

$$-\lambda_0 P_0 + \mu P_1 = 0$$

$$\lambda_0 P_0 - (\lambda_1 + \mu)P_1 + \mu P_2 = 0$$

$$\lambda_1 P_1 - (\lambda_2 + \mu)P_2 + \mu P_3 = 0$$

$$\lambda_2 P_2 - (\lambda_3 + \mu)P_3 + \mu P_4 = 0$$

$$\lambda_3 P_3 - (\lambda_4 + \mu)P_4 + \mu P_5 = 0$$

$$\lambda_4 P_4 - \mu P_5 = 0$$

Solving these equations recursively gives

$$P_1 = \frac{\lambda_0}{\mu}P_0$$

$$P_2 = \frac{\lambda_0\lambda_1}{\mu^2}P_0$$

$$P_3 = \frac{\lambda_0\lambda_1\lambda_2}{\mu^3}P_0$$

$$P_4 = \frac{\lambda_0\lambda_1\lambda_2\lambda_3}{\mu^4}P_0$$

$$P_5 = \frac{\lambda_0\lambda_1\lambda_2\lambda_3\lambda_4}{\mu^5}P_0$$

(a) The probability that the system is empty = P_0. As before $\sum_{n=0}^{5} P_n = 1$ and $\lambda_n = (5-n)\lambda$, therefore

$$P_0\left(1 + \frac{\lambda_0}{\mu} + \frac{\lambda_0\lambda_1}{\mu^2} + \frac{\lambda_0\lambda_1\lambda_2}{\mu^3} + \frac{\lambda_0\lambda_1\lambda_2\lambda_3}{\mu^4} + \frac{\lambda_0\lambda_1\lambda_2\lambda_3\lambda_4}{\mu^5}\right) = 1$$

$$P_0\left[1 + 5\frac{\lambda}{\mu} + 5 \times 4\left(\frac{\lambda}{\mu}\right)^2 + 5 \times 4 \times 3\left(\frac{\lambda}{\mu}\right)^3 + 5 \times 4 \times 3 \times 2\left(\frac{\lambda}{\mu}\right)^4 + 5 \times 4 \times 3 \times 2 \times 1\left(\frac{\lambda}{\mu}\right)^5\right] = 1$$

λ = 1 per hour, μ = 10 per hour, therefore $\rho = \lambda/\mu = 1/10$.

Thus

$$P_0(1 + 0.5 + 0.2 + 0.06 + 0.013 + 0.0012) = 1 = P_0 1.7732$$

$$P_0 = \frac{1}{1.7732} = 0.564$$

$$P_1 = 5\rho \times 0.564 = 0.282$$

$$P_2 = 4\rho \times 0.282 = 0.113$$

$$P_3 = 3\rho \times 0.113 = 0.0.34$$

$$P_4 = 2\rho \times 0.034 = 0.0068$$

$$P_5 = \rho \times 0.0068 = 0.0007$$

(b) To find the average delay for any machine waiting for the start of resetting.

(i) Short Method Based on $\bar{w} = \bar{q}/\bar{\lambda}$ Relationship

n	q	P_n	qP_n	λ_n	$\lambda_n P_n$
0	0	0.564	0.000	5λ	2.820λ
1	0	0.282	0.000	4λ	1.128λ
2	1	0.113	0.113	3λ	0.339λ
3	2	0.034	0.068	2λ	0.068λ
4	3	0.0068	0.020	λ	0.007λ
5	4	0.0007	0.003	0	0.0000
		$\bar{q}$ =	0.204	$\bar{\lambda}$ =	4.362λ

Therefore waiting time in queue is

$$\bar{w} = \frac{\bar{q}}{\bar{\lambda}} = \frac{0.204}{4.362} \times 60 = 2.8 \text{ min}$$

(ii) Short Method Based on $\bar{w} = \bar{n}/\bar{\mu}$ Relationship

Any machine arriving finds n in the system. When all n have been serviced then the new machine will have finished queueing and just be starting service. $\bar{n} = \sum_{n=0}^{5} nP_n$ but the machines arriving depend upon P_n, therefore

$$\bar{w} = (1/\mu) \sum_{n=0}^{5} (nP_n \lambda_n/\bar{\lambda})$$

and $1/\mu$ = 6 min.

n	P_n	λ_n	$\lambda_n/\bar{\lambda}$	nP_n	$nP_n\lambda_n/\bar{\lambda}$
0	.564	5λ		0.000	0.000
1	.282	4λ	0.917	0.282	0.259
2	.113	3λ	0.688	0.226	0.155
3	.034	2λ	0.459	0.102	0.047
4	.0068	1λ	0.229	0.027	0.006
5	.0007	0	0.000	0.003	0.000
			$\bar{n}$ =	0.640	0.467

Therefore $\bar{w} = 6 \times 0.467 = 2.8$ min.

(iii) Full Method Based on Waiting Time Distribution

Assume any machine arriving finds n in the system. The new arrival will wait until all the n machines have been served (in time t+dt). Then the waiting time distribution w(t)dt is based on

(a) probability of an arrival (depends on system size)
(b) system size on arrival
(c) n - 1 machines being served in time t
(d) the nth machine being served in interval (t+dt)-t

Now $\lambda_n = (5 - n)\lambda$ and $\bar{\lambda} = \sum_{n=0}^{5} \lambda_n P_n = (5 - \bar{n})\lambda = 4.36\lambda$. So

$$w(t)\,dt = \sum_{n=1}^{5} (\lambda_n/\bar{\lambda})\, P_n \frac{(\mu t)^{n-1}}{(n-1)!} e^{-\mu t} \mu dt$$

$$= \sum_{n=1}^{5} \frac{(5-n)}{4.36} P_n \frac{\mu^n}{(n-1)!} t^{n-1} e^{-\mu t} dt$$

but $\bar{w} = w(t)t\ dt$

$$= \sum_{n=1}^{5} \frac{(5-n)}{4.36} P_n \frac{\mu^n}{(n-1)!} \int_0^\infty t^n e^{-\mu t} dt$$

Integrating term by term

$$n = 1, \int_0^\infty te^{-\mu t} dt = \frac{1}{\mu} \int_0^\infty e^{-\mu t} dt = \frac{1}{\mu}$$

$$n = 2, \int_0^\infty t^2 e^{-\mu t} dt = \frac{2}{\mu} \int_0^\infty te^{-\mu t} dt = \frac{2}{\mu^3}$$

$$n = 3, \int_0^\infty t^3 e^{-\mu t} dt = \frac{3}{\mu} \int_0^\infty t^2 e^{-\mu t} dt = \frac{6}{\mu^4}$$

$$n = 4, \int_0^\infty t^4 e^{-\mu t} dt = \frac{4}{\mu} \int_0^\infty t^3 e^{-\mu t} dt = \frac{24}{\mu^5}$$

generally $\int_0^\infty t^n e^{-\mu t} dt = \frac{n!}{\mu^{n+1}}$

therefore

$$\bar{w} = \sum_{n=1}^{5} \frac{5-n}{4.36} P_n \frac{\mu^n}{(n-1)!} \times \frac{n!}{\mu^{n+1}} = \sum_{n=1}^{4} \frac{5-n}{4.36} P_n \frac{n}{\mu}$$

This has now reduced to the format in *(ii)* above.

2. The steady-state equations are

$$-\lambda_0 P_0 + \mu_1 P_1 = 0$$

$$\lambda_{n-1} P_{n-1} - (\lambda_n + \mu_n) P_n + \mu_{n+1} P_{n+1} = 0 \qquad n > 0$$

Solving recursively

$$P_1 = \frac{\lambda_0}{\mu_1} P_0$$

$$P_2 = \frac{\lambda_0 \lambda_1}{\mu_1 \mu_2} P_0$$

$$P_n = \frac{\lambda_0 \lambda_1 \ldots \lambda_{n-1}}{\mu_1 \mu_2 \ldots \mu_n} P_0$$

where P_0 is given by $\sum_0^\infty P_n = 1$, that is

$$P_0 = \left(1 + \frac{\lambda_0}{\mu_1} + \frac{\lambda_0\lambda_1}{\mu_1\mu_2} + \quad \ldots \quad + \frac{\lambda_0\lambda_1 \ldots \lambda_{n-1}}{\mu_1\mu_2 \ldots \mu_n} + \ldots\right)^{-1}$$

(a) $\lambda_n = \lambda$; $\mu_n = n\mu$. Arrival rate independent of number in system with parameter λ. Service rate varies according to number in system. This model describes a system with ample servers so that a queue never forms.

(b) $\lambda_n = \lambda$, $\mu_n = n\mu$ $(n \leqslant c)$, $\mu_n = c\mu$ $(n \geqslant c)$. Arrival rate independent of number in system with parameter λ. c servers so that for $n \leqslant c$ no queue forms and the service rate has parameter $n\mu$. For $n > c$, all the servers are busy and a queue of n-c customers is formed.

(c) $\lambda_n = \lambda/(n+1)$, $\mu_n = \mu$. Arrival rate dependent on number in system, service rate independent with parameter μ. This describes a queueing system with discouragement, that is, the sight of a long queue discourages fresh customers from joining it.

(d) $\lambda_n = \lambda (n \leqslant k - 1)$, $\lambda_n = 0$ $(n \geqslant k)$; $\mu_n = \mu$. Arrival rate dependent on queue size, service rate independent with parameter μ. This describes a queue with limited waiting room (k only in system). Thus customers who arrive to find the system full go elsewhere for service.

(e) $\lambda_n = \lambda (n \leqslant k - 1)$, $\lambda_n = 0$ $(n \geqslant k)$, $\mu_n = n\mu$. This describes a system with k servers and waiting room for only k customers, for example a telephone exchange with just k lines and no facility for holding subscribers who require a line but cannot be supplied with one. It is assumed that such calls are lost.

For case (e)

$$P_n = \frac{\lambda^n}{\mu^n \, n!} P_0$$

$$= \frac{\rho^n}{n!} P_0$$

where

$$P_0 \left(1 + \rho + \frac{\rho^2}{2!} + \ldots + \frac{\rho^k}{k!}\right) = 1$$

$$P_n = \frac{\frac{1}{n!}\rho^n}{1 + \rho + \frac{\rho^2}{2!} + \ldots + \frac{\rho^k}{k!}}$$

3. If $n \leqslant N$, $N-n$ spares are on order. Thus

$$\lambda_n = (N-n)\lambda \qquad n \geqslant N$$

and $\lambda_n = 0 \qquad n \geqslant N$

Since the maximum inventory is N

one completion of service = one item used

Probability (1 completion of service in $t, t+\delta t$) = $\mu\delta t$

where

the difference differential equations are

$$P_0(t+\delta t) = P_0(t)(1-\lambda_0\delta t) + P_1(t)\mu\delta t$$

$$P_n(t+\delta t) = P_{n-1}(t)\lambda_{n-1}\delta t + P_n(t)\left[1-(\lambda_n+\mu)\delta t\right]$$

$$+ P_{n+1}(t)\mu\delta t \qquad 0 < n < N$$

$$P_N(t+\delta t) = P_{N-1}(t)\lambda_{N-1}\delta t + P_N(t)(1-\mu\delta t) \qquad n = N$$

In the steady state

$$P_n(t+\delta t) = P_n(t)$$

Thus the steady-state equations are

$$-\lambda_0 P_0 + \mu P_1 = 0 \qquad n = 0$$

$$\lambda_{n-1}P_{n-1} - (\lambda_n+\mu)\, P_n + \mu P_{n+1} = 0 \qquad 0 < n < N$$

$$-\mu P_N + \lambda_{N-1}\, P_{N-1} = 0 \qquad n = N$$

thus $P_1 = \frac{\lambda_0}{\mu} P_0$

$$P_{n+1} = \frac{-\lambda_{n-1}}{\mu} P_{n-1} + \frac{(\lambda_n+\mu)}{\mu} P_n \qquad (6.1)$$

$$P_N = \frac{\lambda_{N-1}}{\mu} P_{N-1}$$

From equation 6.1, with $n = 1$

$$P_2 = -\frac{\lambda_0}{\mu} P_0 + \left(\frac{\lambda_1}{\mu} + 1\right)P_1$$

$$= -P_1 + \left(\frac{\lambda_1}{\mu} + 1\right)P_1 = \frac{\lambda_1}{\mu} P_1 = \frac{\lambda_0\lambda_1}{\mu^2} P_0$$

By recursion

$$P_n = P_0 \frac{\lambda_0\lambda_1\lambda_2 \ldots \lambda_{n-1}}{\mu^n} = \frac{N!}{(N-n)!}\left(\frac{\lambda}{\mu}\right)^n P_0$$

$$\sum_{n=0}^{N} P_n = 1$$

$$P_0 \sum_{n=0}^{N} \frac{N!}{(N-n)!} \frac{\lambda^n}{\mu^n} = 1$$

thus $P_0 = \left[\sum_{n=0}^{N} \frac{N!}{(N-n)!} \frac{\lambda^n}{\mu^n}\right]^{-1}$

Total cost C_T = average stock × stockholding cost + premium × average usage × probability of no stock

$$= £\frac{1}{2} \times \frac{1}{100} \times 100 \sum_{n=1}^{N} nP_n + 15 \times 2P_0$$

$$= £\frac{1}{2} \sum_{n=0}^{N} nP_n + 30P_0$$

We require the value of N that minimises this total cost.

	N				
	1	2	3	4	5
P_0	1/2	1/5	1/16	1/65	1/326
P_1	1/2	2/5	3/16	4/65	5/326
P_2		2/5	6/16	12/65	20/326
P_3				24/65	60/326
P_4				24/65	120/326
P_5					120/326
C_T	15.25	6.6	2.9	2.0	4.2

Thus N = 4 gives maximum stock level for minimum cost.

4. Here $\lambda_n = \lambda/(n+1)$ $\qquad \mu_n = \mu$

Thus

$$P_n = \frac{\lambda_0 \cdots \lambda_{n-1}}{\mu_1 \cdots \mu_n} P_0$$

$$= \left(\frac{\lambda}{\mu}\right)^n \frac{1}{n!} P_0$$

$$= \rho^n \frac{1}{n!} P_0$$

$$P_0 \left(1 + \rho + \frac{\rho^2}{2!} + \frac{\rho^3}{3!} + + \cdots\right) = 1$$

$$P_0 = e^{-\rho}$$

The average number in the queue is given by

$$\bar{q} = \sum_{n=1}^{\infty} (n-1) P_n$$

$$= e^{-\rho}\left(\frac{\rho^2}{2!} + 2\frac{\rho^3}{3!} + 3\frac{\rho^4}{4!} + \cdots\right)$$

$$= e^{-\rho} S$$

Now $e^{\rho} = 1 + \rho + \frac{\rho^2}{2!} + \frac{\rho^3}{3!} + \cdots$

therefore

$$\frac{e^{\rho}}{\rho} = \frac{1}{\rho} + 1 + \frac{\rho}{2!} + \frac{\rho^2}{3!} + \cdots$$

$$\frac{d}{d\rho}\left(\frac{e^{\rho}}{\rho}\right) = \frac{-1}{\rho^2} + \frac{1}{2!} + \frac{2}{3!}\rho + \frac{3}{4!}\rho^2 + \cdots$$

$$\rho^2 \frac{d}{d\rho}\left(\frac{1}{\rho} e^{\rho}\right) = -1 + \frac{\rho^2}{2!} + \frac{2\rho^3}{3!} + \frac{3\rho^4}{4!} + \cdots$$

thus $\rho^2\left(\frac{-1}{\rho^2} e^{\rho} + \frac{1}{\rho} e^{\rho}\right) = -1 + S$

and $S = 1 - e^{\rho} + \rho e^{\rho}$

$$\bar{q} = e^{-\rho} S = e^{-\rho} - 1 + \rho = e^{-\lambda/\mu} - 1 + \frac{\lambda}{\mu}$$

5. For this system, $\lambda = 3$/day, $\mu_n = n\mu$/day, where μ = number of ships that may be unloaded per day with one crew $\mu = 2$. Let $\rho = \lambda/\mu = 1.5$. Then steady-state equations are

$$-\rho P_0 + P_1 = 0 \qquad n = 0$$

$$\rho P_{n-1} - (n + \rho)P_n + (n + 1)P_{n+1} = 0 \qquad n > 0$$

Solving recursively gives

$$P_n = \frac{\rho^n}{n!}P_0$$

$$P_0\left(1 + \rho + \frac{\rho^2}{2!} + \frac{\rho^3}{3!} + \ldots\right) = 1$$

$$P_0 = e^{-\rho}$$

Average number of unloading crews = average number of ships in system, therefore

$$\bar{n} = \sum_{n=0}^{\infty} nP_n$$

$$= e^{-\rho}\left(\rho + \frac{2\rho^2}{2!} + \frac{3\rho^3}{3!} + \ldots\right)$$

$$= \rho = 1.5 \text{ crews}$$

$$\text{Prob. (more than 4 crews)} = \sum_{n=5}^{\infty} P_n = 1 - \sum_{n=0}^{4} P_n = .019$$

7 SINGLE-CHANNEL SYSTEMS WITH GENERAL SERVICE TIME DISTRIBUTIONS (M/G/1/ SYSTEMS)

7.1 INTRODUCTION

The general formulae for these systems were obtained by Pollaczeh-Khintchine, while Fry obtained a solution to systems with constant service times, that is, M/D/1/∞ systems.

7.2 RÉSUMÉ OF BASIC THEORY

The formulae of Pollaczeh-Khintchine are general and also cover the M/D/1/∞ system that Fry studied.

Probability of no customers in system $P_0 = 1 - \rho$

(For P_n for M/D/1 systems only see Fry.)

Average queue length $\bar{q} = \rho^2 \dfrac{(1+\mu^2\sigma_s^2)}{2(1-\rho)}$

Average number in system $\bar{n} = \bar{q} + \rho$

Average waiting time $\bar{w} = \dfrac{\rho(1+\mu^2\sigma_s^2)}{2\mu(1-\rho)}$

Average time in system $\bar{d} = \bar{w} + \dfrac{1}{\mu}$

7.3 PROBLEMS

1. The time to process a claim in an insurance office is exactly 25 minutes for each claim submitted. If claimants arrive in random stream at the average rate of one every 40 minutes, how long on average must a claimant wait for service given that there is only one insurance clerk processing the claims.

2. In a heavy machine shop, the overhead crane is 75 per cent utilised. Time study observations gave the average slinging time as 10.5 min with a standard deviation of 8.8 min. What is the average calling rate for the services of the crane, and what is the average delay in getting service? If the average service time is cut to 8.0 min, with standard deviation of 6.0 min, how much reduction will occur, on average, in the delay of getting served?

3. A firm employs a team of skilled fitters to repair breakdowns that occur to its machines. Consider the following alternative methods of operation.

(a) No specialisation, all fitters tackle all repairs. This method gives a standard deviation of repair time equal to the mean repair rate (coefficient of variation = 1.0)

(b) Partial specialisation, which reduces the standard deviation of repair time to half the mean repair rate

(c) Full specialisation, which ensures that all repairs are completed in the same time.

Given that the mean repair rate is the same for all methods of operation and that repairs arrive at random with respect to time show that

(i) Method (b) gives a 37.5 per cent reduction on average waiting time for repair over method (a)

(ii) Method (c) gives a 50 per cent reduction on average waiting time over method (a).

7.4 SOLUTIONS

1. Here $\sigma_s^2 = 0$, $\lambda = 1.5/h$, $\mu = 2.4/h$, $\rho = 1.5/2.4 = 0.625$

Average waiting time is $\bar{w} = \rho \dfrac{(1+\mu^2\sigma_s^2)}{2\mu(1-\rho)} = \dfrac{\rho}{2(1-\rho)\mu}$

$$= \frac{0.625}{2(1-0.625)\times 2.4}$$

$$= 0.35 \text{ h}$$

$$= 21 \text{ min}$$

2. This is a $M/G/1/\infty$ process. Thus the average delay in getting service $\bar{w}$ is given by

$$\bar{w} = \frac{\rho}{2\mu(1-\rho)}(1 + \mu^2\sigma_s^2)$$

Initial situation

$$\rho = 0.75$$

$$\mu = \frac{60}{10.5} = 5.71/h$$

$$\lambda = \rho \times \mu$$

$$= 0.75 \times 5.71 = 4.29/h$$

Average waiting time is

$$\bar{w} = \frac{\rho}{2(1-\rho)} (1 + \mu^2\sigma_s^2) \times \frac{1}{\mu}$$

$$\bar{w} = \frac{0.75}{2(1-0.75)} \left(1 + 5.71^2 \times \frac{8.8^2}{60^2}\right) \times \frac{60}{5.71} \text{ min}$$

$$\bar{w} = \frac{0.75}{0.5} \times 1.70 \times \frac{60}{5.71} \text{ min}$$

$$\bar{w} = 26.8 \text{ min}$$

If service time is cut to 8 minutes

$$\mu = \frac{60}{8} = 7.5/\text{h}$$

$$\rho = \frac{4.29}{7.5} = 0.571$$

or utilisation of the crane reduced to 57.1 per cent

$$\bar{w} = \frac{0.571}{2(1-0.571)} \left(1 + 7.5^2 \times \frac{6.0^2}{60^2}\right) \times \frac{60}{7.5} \text{ min}$$

$$= \frac{0.571}{2\times 0.429} \times 1.562 \times 8$$

$$= 8.3 \text{ min}$$

a reduction of 18.5 min or approximately 70 per cent.

3. All these are $M/G/1/\infty$ processes with equal service rate μ and arrival rate λ. The average waiting time for an $M/G/1/\infty$ process is given by

$$\bar{w} = \frac{\rho}{2\mu(1-\rho)} (1 + \mu^2\sigma_s^2)$$

where σ_s^2 is the variance of the service time. Therefore

(a) $\sigma_s^2 = 1/\mu^2$

$$\bar{w}_a = \frac{\rho}{\mu(1-\rho)}$$

(b) $\sigma_s^2 = 1/4\mu^2$

$$\bar{w}_b = \frac{\rho\left(1 + \mu^2 \frac{1}{4\mu^2}\right)}{2\mu(1-\rho)} = \frac{5\rho}{8\mu(1-\rho)}$$

(c) $\sigma_s^2 = 0$

$$\bar{w}_c = \frac{\rho}{2\mu(1-\rho)}$$

The cost arising from the waiting in the queue for jobs is proportional to $\lambda\bar{w}$. The percentage savings achieved are

(i) saving in waiting time

$$= \frac{w_a - w_b}{wa} \times 100 \text{ per cent}$$

$$= \frac{\frac{\rho}{\mu(1-\rho)} - \frac{5}{8}\frac{\rho}{\mu(1-\rho)}}{\frac{\rho}{\mu(1-\rho)}} \times 100 \text{ per cent}$$

$$= 300/8 = 37.5 \text{ per cent}$$

(ii) % saving in waiting time

$$= \frac{w_a - w_c}{wa} \times 100 \text{ per cent}$$

$$= \frac{\frac{\rho}{\mu(1-\rho)} - \frac{\rho}{2\mu(1-\rho)}}{\frac{\rho}{\mu(1-\rho)}} \times 100 \text{ per cent} = 50 \text{ per cent}$$

STATISTICAL TABLES

TABLE 1 POISSON DISTRIBUTION

The table gives the probability that r or more random events are contained in an interval when the average number of such events per interval is m, that is

$$\sum_{x=r}^{\infty} e^{-m} \frac{m^x}{x!}$$

Where there is no entry for a particular pair of values of r and m, this indicates that the appropriate probability is less than 0.000 05. Similarly, except for the case r = 0 when the entry is exact, a tabulated value of 1.0000 represents a probability greater than 0.999 95.

m =	0.1	0.2	0.3	0.4	0.5	0.6	0.7	0.8	0.9	1.0
r = 0	1.0000	1.0000	1.0000	1.0000	1.0000	1.0000	1.0000	1.0000	1.0000	1.0000
1	.0952	.1813	.2592	.3297	.3935	.4512	.5034	.5507	.5934	.6321
2	.0047	.0175	.0369	.0616	.0902	.1219	.1558	.1912	.2275	.2642
3	.0002	.0011	.0036	.0079	.0144	.0231	.0341	.0474	.0629	.0803
4		.0001	.0003	.0008	.0018	0034	.0058	.0091	.0135	.0190
5				.0001	.0002	.0004	.0008	.0014	.0023	.0037
6							.0001	.0002	.0003	.0006
7										.0001

m =	1.1	1.2	1.3	1.4	1.5	1.6	1.7	1.8	1.9	2.0
r = 0	1.0000	1.0000	1.0000	1.0000	1.0000	1.0000	1.0000	1.0000	1.0000	1.0000
1	.6671	.6988	.7275	.7534	.7769	.7981	.8173	.8347	.8504	.8647
2	.3010	.3374	.3732	.4082	.4422	.4751	.5068	.5372	.5663	.5940
3	.0996	.1205	.1429	.1665	.1912	.2166	.2428	.2694	.2963	.3233
4	.0257	.0338	.0431	.0537	.0656	.0788	.0932	.1087	.1253	.1429
5	.0054	.0077	.0107	.0143	.0186	.0237	.0296	.0364	.0441	.0527
6	.0010	.0015	.0022	.0032	.0045	.0060	.0080	.0104	.0132	.0166
7	.0001	.0003	.0004	.0006	.0009	.0013	.0019	.0026	.0034	.0045
8			.0001	.0001	.0002	.0003	.0004	.0006	.0008	.0011
9							.0001	.0001	.0002	.0002

m =	2.1	2.2	2.3	2.4	2.5	2.6	2.7	2.8	2.9	3.0
r = 0	1.0000	1.0000	1.0000	1.0000	1.0000	1.0000	1.0000	1.0000	1.0000	1.0000
1	.8775	.8892	.8997	.9093	.9179	.9257	.9328	.9392	.9450	.9502
2	.6204	.6454	.6691	.6916	.7127	.7326	.7513	.7689	.7854	.8009
3	.3504	.3773	.4040	.4303	.4562	.4816	.5064	.5305	.5540	.5768
4	.1614	.1806	.2007	.2213	.2424	.2640	.2859	.3081	.3304	.3528
5	.0621	.0725	.0838	.0959	.1088	.1226	.1371	.1523	.1682	.1847
6	.0204	.0249	.0300	.0357	.0420	.0490	.0567	.0651	.0742	.0839
7	.0059	.0075	.0094	.0116	.0142	.0172	.0206	.0244	.0287	.0335
8	.0015	.0020	.0026	.0033	.0042	.0053	.0066	.0081	.0099	.0119
9	.0003	.0005	.0006	.0009	.0011	.0015	.0019	.0024	.0031	.0038
10	.0001	.0001	.0001	.0002	.0003	.0004	.0005	.0007	.0009	.0011
11					.0001	.0001	.0001	.0002	.0002	.0003
12									.0001	.0001

m =	3.1	3.2	3.3	3.4	3.5	3.6	3.7	3.8	3.9	4.0
r = 0	1.0000	1.0000	1.0000	1.0000	1.0000	1.0000	1.0000	1.0000	1.0000	1.0000
1	.9550	.9592	.9631	.9666	.9698	.9727	.9753	.9776	.9798	.9817
2	.8153	.8288	.8414	.8532	.8641	.8743	.8838	.8926	.9008	.9084
3	.5988	.6201	.6406	.6603	.6792	.6973	.7146	.7311	.7469	.7619
4	.3752	.3975	.4197	.4416	.4634	.4848	.5058	.5265	.5468	.5665
5	.2018	.2194	.2374	.2558	.2746	.2936	.3128	.3322	.3516	.3712
6	.0943	.1054	.1171	.1295	.1424	.1559	.1699	.1844	.1994	.2149
7	.0388	.0446	.0510	.0579	.0653	.0733	.0818	.0909	.1005	.1107
8	.0142	.0168	.0198	.0231	.0267	.0308	.0352	.0401	.0454	.0511
9	.0047	.0057	.0069	.0083	.0099	.0117	.0137	.0160	.0185	.0214
10	.0014	.0018	.0022	.0027	.0033	.0040	.0048	.0058	.0069	.0081
11	.0004	.0005	.0006	.0008	.0010	.0013	.0016	.0019	.0023	.0028
12	.0001	.0001	.0002	.0002	.0003	.0004	.0005	.0006	.0007	.0009
13				.0001	.0001	.0001	.0001	.0002	.0002	.0003
14									.0001	.0001

m =	4.1	4.2	4.3	4.4	4.5	4.6	4.7	4.8	4.9	5.0
r = 0	1.0000	1.0000	1.0000	1.0000	1.0000	1.0000	1.0000	1.0000	1.0000	1.0000
1	.9834	.9850	.9864	.9877	.9889	.9899	.9909	.9918	.9926	.9933
2	.9155	.9220	.9281	.9337	.9389	.9437	.9482	.9523	.9561	.9596
3	.7762	.7898	.8026	.8149	.8264	.8374	.8477	.8575	.8667	.8753
4	.5858	.6046	.6228	.6406	.6577	.6743	.6903	.7058	.7207	.7350
5	.3907	.4102	.4296	.4488	.4679	.4868	.5054	.5237	.5418	.5595
6	.2307	.2469	.2633	.2801	.2971	.3142	.3316	.3490	.3665	.3840
7	.1214	.1325	.1442	.1564	.1689	.1820	.1954	.2092	.2233	.2378
8	.0573	.0639	.0710	.0786	.0866	.0951	.1040	.1133	.1231	.1334
9	.0245	.0279	.0317	.0358	.0403	.0451	.0503	.0558	.0618	.0681
10	.0095	.0111	.0129	.0149	.0171	.0195	.0222	.0251	.0283	.0318
11	.0034	.0041	.0048	.0057	.0067	.0078	.0090	.0104	.0120	.0137
12	.0011	.0014	.0017	.0020	.0024	.0029	.0034	.0040	.0047	.0055
13	.0003	.0004	.0005	.0007	.0008	.0010	.0012	.0014	.0017	.0020
14	.0001	.0001	.0002	.0002	.0003	.0003	.0004	.0005	.0006	.0007
15				.0001	.0001	.0001	.0001	.0001	.0002	.0002
16									.0001	.0001

m =	5.2	5.4	5.6	5.8	6.0	6.2	6.4	6.6	6.8	7.0
r = 0	1.0000	1.0000	1.0000	1.0000	1.0000	1.0000	1.0000	1.0000	1.0000	1.0000
1	.9945	.9955	.9963	.9970	.9975	.9980	.9983	.9986	.9989	.9991
2	.9658	.9711	.9756	.9794	.9826	.9854	.9877	.9897	.9913	.9927
3	.8912	.9052	.9176	.9285	.9380	.9464	.9537	.9600	.9656	.9704
4	.7619	.7867	.8094	.8300	.8488	.8658	.8811	.8948	.9072	.9182
5	.5939	.6267	.6579	.6873	.7149	.7408	.7649	.7873	.8080	.8270
6	.4191	.4539	.4881	.5217	.5543	.5859	.6163	.6453	.6730	.6993
7	.2676	.2983	.3297	.3616	.3937	.4258	.4577	.4892	.5201	.5503
8	.1551	.1783	.2030	.2290	.2560	.2840	.3127	.3419	.3715	.4013
9	.0819	.0974	.1143	.1328	.1528	.1741	.1967	.2204	.2452	.2709
10	.0397	.0488	.0591	.0708	.0839	.0984	.1142	.1314	.1498	.1695
11	.0177	.0225	.0282	.0349	.0426	.0514	.0614	.0726	.0849	.0985
12	.0073	.0096	.0125	.0160	.0201	.0250	.0307	.0373	.0448	.0534
13	.0028	.0038	.0051	.0068	.0088	.0113	.0143	.0179	.0221	.0270
14	.0010	.0014	.0020	.0027	.0036	.0048	.0063	.0080	.0102	.0128
15	.0003	.0005	.0007	.0010	.0014	.0019	.0026	.0034	.0044	.0057
16	.0001	.0002	.0002	.0004	.0005	.0007	.0010	.0014	.0018	.0024
17		.0001	.0001	.0001	.0002	.0003	.0004	.0005	.0007	.0010
18					.0001	.0001	.0001	.0002	.0003	.0004
19								.0001	.0001	.0001

m =	7.2	7.4	7.6	7.8	8.0	8.2	8.4	8.6	8.8	9.0
r = 0	1.0000	1.0000	1.0000	1.0000	1.0000	1.0000	1.0000	1.0000	1.0000	1.0000
1	.9993	.9994	.9995	.9996	.9997	.9997	.9998	.9998	.9998	.9999
2	.9939	.9949	.9957	.9964	.9970	.9975	.9979	.9982	.9985	.9988
3	.9745	.9781	.9812	.9839	.9862	.9882	.9900	.9914	.9927	.9938
4	.9281	.9368	.9446	.9515	.9576	.9630	.9677	.9719	.9756	.9788
5	.8445	.8605	.8751	.8883	.9004	.9113	.9211	.9299	.9379	.9450
6	.7241	.7474	.7693	.7897	.8088	.8264	.8427	.8578	.8716	.8843
7	.5796	.6080	.6354	.6616	.6866	.7104	.7330	.7543	.7744	.7932
8	.4311	.4607	.4900	.5188	.5470	.5746	.6013	.6272	.6522	.6761
9	.2973	.3243	.3518	.3796	.4075	.4353	.4631	.4906	.5177	.5443
10	.1904	.2123	.2351	.2589	.2834	.3085	.3341	.3600	.3863	.4126
11	.1133	.1293	.1465	.1648	.1841	.2045	.2257	.2478	.2706	.2940
12	.0629	.0735	.0852	.0980	.1119	.1269	.1429	.1600	.1780	.1970
13	.0327	.0391	.0464	.0546	.0638	.0739	.0850	.0971	.1102	.1242
14	.0159	.0195	.0238	.0286	.0342	.0405	.0476	.0555	.0642	.0739
15	.0073	.0092	.0114	.0141	.0173	.0209	.0251	.0299	.0353	.0415
16	.0031	.0041	.0052	.0066	.0082	.0102	.0125	.0152	.0184	.0220
17	.0013	.0017	.0022	.0029	.0037	.0047	.0059	.0074	.0091	.0111
18	.0005	.0007	.0009	.0012	.0016	.0021	.0027	.0034	.0043	.0053
19	.0002	.0003	.0004	.0005	.0006	.0009	.0011	.0015	.0019	.0024
20	.0001	.0001	.0001	.0002	.0003	.0003	.0005	.0006	.0008	.0011
21				.0001	.0001	.0001	.0002	.0002	.0003	.0004
22							.0001	.0001	.0001	.0002
23										.0001

m =	9.2	9.4	9.6	9.8	10.0	11.0	12.0	13.0	14.0	15.0
r = 0	1.0000	1.0000	1.0000	1.0000	1.0000	1.0000	1.0000	1.0000	1.0000	1.0000
1	.9999	.9999	.9999	.9999	1.0000	1.0000	1.0000	1.0000	1.0000	1.0000
2	.9990	.9991	.9993	.9994	.9995	.9998	.9999	1.0000	1.0000	1.0000
3	.9947	.9955	.9962	.9967	.9972	.9988	.9995	.9998	.9999	1.0000
4	.9816	.9840	.9862	.9880	.9897	.9951	.9977	.9990	.9995	.9998
5	.9514	.9571	.9622	.9667	.9707	.9849	.9924	.9963	.9982	.9991
6	.8959	.9065	.9162	.9250	.9329	.9625	.9797	.9893	.9945	.9972
7	.8108	.8273	.8426	.8567	.8699	.9214	.9542	.9741	.9858	.9924
8	.6990	.7208	.7416	.7612	.7798	.8568	.9105	.9460	.9684	.9820
9	.5704	.5958	.6204	.6442	.6672	.7680	.8450	.9002	.9379	.9626
10	.4389	.4651	.4911	.5168	.5421	.6595	.7576	.8342	.8906	.9301
11	.3180	.3424	.3671	.3920	.4170	.5401	.6528	.7483	.8243	.8815
12	.2168	.2374	.2588	.2807	.3032	.4207	.5384	.6468	.7400	.8152
13	.1393	.1552	.1721	.1899	.2084	.3113	.4240	.5369	.6415	.7324
14	.0844	.0958	.1081	.1214	.1355	.2187	.3185	.4270	.5356	.6368
15	.0483	.0559	.0643	.0735	.0835	.1460	.2280	.3249	.4296	.5343
16	.0262	.0309	.0362	.0421	.0487	.0926	.1556	.2364	.3306	.4319
17	.0135	.0162	.0194	.0230	.0270	.0559	.1013	.1645	.2441	.3359
18	.0066	.0081	.0098	.0119	.0143	.0322	.0630	.1095	.1728	.2511
19	.0031	.0038	.0048	.0059	.0072	.0177	.0374	.0698	.1174	.1805
20	.0014	.0017	.0022	.0028	.0035	.0093	.0213	.0427	.0765	.1248
21	.0006	.0008	.0010	.0012	.0016	.0047	.0116	.0250	.0479	.0830
22	.0002	.0003	.0004	.0005	.0007	.0023	.0061	.0141	.0288	.0531
23	.0001	.0001	.0002	.0002	.0003	.0010	.0030	.0076	.0167	.0327
24			.0001	.0001	.0001	.0005	.0015	.0040	.0093	.0195
25						.0002	.0007	.0020	.0050	.0112
26						.0001	.0003	.0010	.0026	.0062
27							.0001	.0005	.0013	.0033
28							.0001	.0002	.0006	.0017
29								.0001	.0003	.0009
30									.0001	.0004
31									.0001	.0002
32										.0001

m =	16.0	17.0	18.0	19.0	20.0	21.0	22.0	23.0	24.0	25.0
r = 0	1.0000	1.0000	1.0000	1.0000	1.0000	1.0000	1.0000	1.0000	1.0000	1.0000
1	1.0000	1.0000	1.0000	1.0000	1.0000	1.0000	1.0000	1.0000	1.0000	1.0000
2	1.0000	1.0000	1.0000	1.0000	1.0000	1.0000	1.0000	1.0000	1.0000	1.0000
3	1.0000	1.0000	1.0000	1.0000	1.0000	1.0000	1.0000	1.0000	1.0000	1.0000
4	.9999	1.0000	1.0000	1.0000	1.0000	1.0000	1.0000	1.0000	1.0000	1.0000
5	.9996	.9998	.9999	1.0000	1.0000	1.0000	1.0000	1.0000	1.0000	1.0000
6	.9986	.9993	.9997	.9998	.9999	1.0000	1.0000	1.0000	1.0000	1.0000
7	.9960	.9979	.9990	.9995	.9997	.9999	.9999	1.0000	1.0000	1.0000
8	.9900	.9946	.9971	.9985	.9992	.9996	.9998	.9999	1.0000	1.0000
9	.9780	.9874	.9929	.9961	.9979	.9989	.9994	.9997	.9998	.9999
10	.9567	.9739	.9846	.9911	.9950	.9972	.9985	.9992	.9996	.9998
11	.9226	.9509	.9696	.9817	.9892	.9937	.9965	.9980	.9989	.9994
12	.8730	.9153	.9451	.9653	.9786	.9871	.9924	.9956	.9975	.9986
13	.8069	.8650	.9083	.9394	.9610	.9755	.9849	.9909	.9946	.9969
14	.7255	.7991	.8574	.9016	.9339	.9566	.9722	.9826	.9893	.9935
15	.6325	.7192	.7919	.8503	.8951	.9284	.9523	.9689	.9802	.9876
16	.5333	.6285	.7133	.7852	.8435	.8889	.9231	.9480	.9656	.9777
17	.4340	.5323	.6249	.7080	.7789	.8371	.8830	.9179	.9437	.9623
18	.3407	.4360	.5314	.6216	.7030	.7730	.8310	.8772	.9129	.9395
19	.2577	.3450	.4378	.5305	.6186	.6983	.7675	.8252	.8717	.9080
20	.1878	.2637	.3491	.4394	.5297	.6157	.6940	.7623	.8197	.8664
21	.1318	.1945	.2693	.3528	.4409	.5290	.6131	.6899	.7574	.8145
22	.0892	.1385	.2009	.2745	.3563	.4423	.5284	.6106	.6861	.7527
23	.0582	.0953	.1449	.2069	.2794	.3595	.4436	.5277	.6083	.6825
24	.0367	.0633	.1011	.1510	.2125	.2840	.3626	.4449	.5272	.6061
25	.0223	.0406	.0683	.1067	.1568	.2178	.2883	.3654	.4460	.5266
26	.0131	.0252	.0446	.0731	.1122	.1623	.2229	.2923	.3681	.4471
27	.0075	.0152	.0282	.0486	.0779	.1174	.1676	.2277	.2962	.3706
28	.0041	.0088	.0173	.0313	.0525	.0825	.1225	.1726	.2323	.2998
29	.0022	.0050	.0103	.0195	.0343	.0564	.0871	.1274	.1775	.2366
30	.0011	.0027	.0059	.0118	.0218	.0374	.0602	.0915	.1321	.1821
31	.0006	.0014	.0033	.0070	.0135	.0242	.0405	.0640	.0958	.1367
32	.0003	.0007	.0018	.0040	.0081	.0152	.0265	.0436	.0678	.1001
33	.0001	.0004	.0010	.0022	.0047	.0093	.0169	.0289	.0467	.0715
34	.0001	.0002	.0005	.0012	.0027	.0055	.0105	.0187	.0314	.0498
35		.0001	.0002	.0006	.0015	.0032	.0064	.0118	.0206	.0338
36			.0001	.0003	.0008	.0018	.0038	.0073	.0132	.0225
37			.0001	.0002	.0004	.0010	.0022	.0044	.0082	.0146
38				.0001	.0002	.0005	.0012	.0026	.0050	.0092
39					.0001	.0003	.0007	.0015	.0030	.0057
40					.0001	.0001	.0004	.0008	.0017	.0034
41						.0001	.0002	.0004	.0010	.0020
42							.0001	.0002	.0005	.0012
43								.0001	.0003	.0007
44								.0001	.0002	.0004
45									.0001	.0002
46										.0001

m =	26.0	27.0	28.0	29.0	30.0	32.0	34.0	36.0	38.0	40.0
r = 9	1.0000	1.0000	1.0000	1.0000	1.0000	1.0000	1.0000	1.0000	1.0000	1.0000
10	.9999	.9999	1.0000	1.0000	1.0000	1.0000	1.0000	1.0000	1.0000	1.0000
11	.9997	.9998	.9999	1.0000	1.0000	1.0000	1.0000	1.0000	1.0000	1.0000
12	.9992	.9996	.9998	.9999	.9999	1.0000	1.0000	1.0000	1.0000	1.0000
13	.9982	.9990	.9994	.9997	.9998	1.0000	1.0000	1.0000	1.0000	1.0000
14	.9962	.9978	.9987	.9993	.9996	.9999	1.0000	1.0000	1.0000	1.0000
15	.9924	.9954	.9973	.9984	.9991	.9997	.9999	1.0000	1.0000	1.0000
16	.9858	.9912	.9946	.9967	.9981	.9993	.9998	.9999	1.0000	1.0000
17	.9752	.9840	.9899	.9937	.9961	.9986	.9995	.9998	1.0000	1.0000
18	.9580	.9726	.9821	.9885	.9927	.9972	.9990	.9997	.9999	1.0000
19	.9354	.9555	.9700	.9801	.9871	.9948	.9980	.9993	.9998	.9999
20	.9032	.9313	.9522	.9674	.9781	.9907	.9963	.9986	.9995	.9998
21	.8613	.8985	.9273	.9489	.9647	.9841	.9932	.9973	.9990	.9996
22	.8095	.8564	.8940	.9233	.9456	.9740	.9884	.9951	.9981	.9993
23	.7483	.8048	.8517	.8896	.9194	.9594	.9809	.9915	.9965	.9986
24	.6791	.7441	.8002	.8471	.8854	.9390	.9698	.9859	.9938	.9974
25	.6041	.6758	.7401	.7958	.8428	.9119	.9540	.9776	.9897	.9955
26	.5261	.6021	.6728	.7363	.7916	.8772	.9326	.9655	.9834	.9924
27	.4481	.5256	.6003	.6699	.7327	.8344	.9047	.9487	.9741	.9877
28	.3730	.4491	.5251	.5986	.6671	.7838	.8694	.9264	.9611	.9807
29	.3033	.3753	.4500	.5247	.5969	.7259	.8267	.8977	.9435	.9706
30	.2407	.3065	.3774	.4508	.5243	.6620	.7765	.8621	.9204	.9568
31	.1866	.2447	.3097	.3794	.4516	.5939	.7196	.8194	.8911	.9383
32	.1411	.1908	.2485	.3126	.3814	.5235	.6573	.7697	.8552	.9145
33	.1042	.1454	.1949	.2521	.3155	.4532	.5911	.7139	.8125	.8847
34	.0751	.1082	.1495	.1989	.2556	.3850	.5228	.6530	.7635	.8486
35	.0528	.0787	.1121	.1535	.2027	.3208	.4546	.5885	.7086	.8061
36	.0363	.0559	.0822	.1159	.1574	.2621	.3883	.5222	.6490	.7576
37	.0244	.0388	.0589	.0856	.1196	.2099	.3256	.4558	.5862	.7037
38	.0160	.0263	.0413	.0619	.0890	.1648	.2681	.3913	.5216	.6453
39	.0103	.0175	.0283	.0438	.0648	.1268	.2166	.3301	.4570	.5840
40	.0064	.0113	.0190	.0303	.0463	.0956	.1717	.2737	.3941	.5210
41	.0039	.0072	.0125	.0205	.0323	.0707	.1336	.2229	.3343	.4581
42	.0024	.0045	.0080	.0136	.0221	.0512	.1019	.1783	.2789	.3967
43	.0014	.0027	.0050	.0089	.0148	.0364	.0763	.1401	.2288	.3382
44	.0008	.0016	.0031	.0056	.0097	.0253	.0561	.1081	.1845	.2838
45	.0004	.0009	.0019	.0035	.0063	.0173	.0404	.0819	.1462	.2343
46	.0002	.0005	.0011	.0022	.0040	.0116	.0286	.0609	.1139	.1903
47	.0001	.0003	.0006	.0013	.0025	.0076	.0199	.0445	.0872	.1521
48	.0001	.0002	.0004	.0008	.0015	.0049	.0136	.0320	.0657	.1196
49		.0001	.0002	.0004	.0009	.0031	.0091	.0225	.0486	.0925
50			.0001	.0002	.0005	.0019	.0060	.0156	.0353	.0703
51			.0001	.0001	.0003	.0012	.0039	.0106	.0253	.0526
52				.0001	.0002	.0007	.0024	.0071	.0178	.0387
53					.0001	.0004	.0015	.0047	.0123	.0281
54					.0001	.0002	.0009	.0030	.0084	.0200
55						.0001	.0006	.0019	.0056	.0140
56						.0001	.0003	.0012	.0037	.0097
57							.0002	.0007	.0024	.0066
58							.0001	.0005	.0015	.0044
59							.0001	.0003	.0010	.0029
60								.0002	.0006	.0019
61								.0001	.0004	.0012
62								.0001	.0002	.0008
63									.0001	.0005
64									.0001	.0003
65										.0002
66										.0001
67										.0001

For values of m greater than 40 use the Normal Distribution curve by setting $\mu = m$ and $\sigma = \sqrt{m}$.

(See J. Murdoch and J.A. Barnes, Statistical Tables, Macmillan, London and Basingstoke, 1971.)

TABLE 2 EXPONENTIAL FUNCTION e^{-x}

For any negative exponential distribution, the tabulated function may be used to find the proportion of the distribution in excess of x times the mean. As an example, in random sampling of an exponential variate with a mean of 8, the probability that a single value will exceed 6 is 0.4724. Further, the 1% point of the distribution is seen to be 4.61 times the mean.

x	.0	.1	.2	.3	.4	.5	.6	.7	.8	.9
1.0	.3679	.3329	.3012	.2725	.2466	.2231	.2019	.1827	.1653	.1496
2.0	.1353	.1225	.1108	.1003	.0907	.0821	.0743	.0672	.0608	.0550
3.0	.0498	.0450	.0408	.0369	.0334	.0302	.0273	.0247	.0224	.0202
4.0	.0183	.0166	.0150	.0136	.0123	.0111	.0101	$.0^{2}910$	$.0^{2}823$	$.0^{2}745$
5.0	$.0^{2}674$	$.0^{2}610$	$.0^{2}552$	$.0^{2}499$	$.0^{2}452$	$.0^{2}409$	$.0^{2}370$	$.0^{2}335$	$.0^{2}303$	$.0^{2}274$
6.0	$.0^{2}248$	$.0^{2}224$	$.0^{2}203$	$.0^{2}184$	$.0^{2}166$	$.0^{2}150$	$.0^{2}136$	$.0^{2}123$	$.0^{2}111$	$.0^{2}101$
7.0	$.0^{3}912$	$.0^{3}825$	$.0^{3}747$	$.0^{3}676$	$.0^{3}611$	$.0^{3}553$	$.0^{3}500$	$.0^{3}453$	$.0^{3}410$	$.0^{3}371$
8.0	$.0^{3}335$	$.0^{3}304$	$.0^{3}275$	$.0^{3}249$	$.0^{3}225$	$.0^{3}203$	$.0^{3}184$	$.0^{3}167$	$.0^{3}151$	$.0^{3}136$
9.0	$.0^{3}123$	$.0^{3}112$	$.0^{3}101$	$.0^{4}914$	$.0^{4}827$	$.0^{4}749$	$.0^{4}677$	$.0^{4}613$	$.0^{4}555$	$.0^{4}502$
10.0	$.0^{4}454$	$.0^{4}411$	$.0^{4}372$	$.0^{4}336$	$.0^{4}304$	$.0^{4}275$	$.0^{4}249$	$.0^{4}225$	$.0^{4}204$	$.0^{4}185$
11.0	$.0^{4}167$	$.0^{4}151$	$.0^{4}137$	$.0^{4}124$	$.0^{4}112$	$.0^{4}101$	$.0^{5}917$	$.0^{5}829$	$.0^{5}750$	$.0^{5}679$
12.0	$.0^{5}614$	$.0^{5}556$	$.0^{5}503$	$.0^{5}455$	$.0^{5}412$	$.0^{5}373$	$.0^{5}337$	$.0^{5}305$	$.0^{5}276$	$.0^{5}250$
13.0	$.0^{5}226$	$.0^{5}205$	$.0^{5}185$	$.0^{5}167$	$.0^{5}152$	$.0^{5}137$	$.0^{5}124$	$.0^{5}112$	$.0^{5}102$	$.0^{6}919$
14.0	$.0^{6}832$	$.0^{6}752$	$.0^{6}681$	$.0^{6}616$	$.0^{6}557$	$.0^{6}504$	$.0^{6}456$	$.0^{6}413$	$.0^{6}374$	$.0^{6}338$
15.0	$.0^{6}306$	$.0^{6}277$	$.0^{6}250$	$.0^{6}227$	$.0^{6}205$	$.0^{6}186$	$.0^{6}168$	$.0^{6}152$	$.0^{6}137$	$.0^{6}124$
16.0	$.0^{6}113$	$.0^{6}102$	$.0^{7}921$	$.0^{7}834$	$.0^{7}754$	$.0^{7}683$	$.0^{7}618$	$.0^{7}559$	$.0^{7}506$	$.0^{7}458$
17.0	$.0^{7}414$	$.0^{7}375$	$.0^{7}339$	$.0^{7}307$	$.0^{7}278$	$.0^{7}251$	$.0^{7}227$	$.0^{7}206$	$.0^{7}186$	$.0^{7}168$
18.0	$.0^{7}152$	$.0^{7}138$	$.0^{7}125$	$.0^{7}113$	$.0^{7}102$	$.0^{8}924$	$.0^{8}836$	$.0^{8}756$	$.0^{8}684$	$.0^{8}619$
19.0	$.0^{8}560$	$.0^{8}507$	$.0^{8}459$	$.0^{8}415$	$.0^{8}376$	$.0^{8}340$	$.0^{8}307$	$.0^{8}278$	$.0^{8}252$	$.0^{8}228$
20.0	$.0^{8}206$									

x	.00	.01	.02	.03	.04	.05	.06	.07	.08	.09
0	1.0000	.9900	.9802	.9704	.9608	.9512	.9418	.9324	.9231	.9139
.1	.9048	.8958	.8869	.8781	.8694	.8607	.8521	.8437	.8353	.8270
.2	.8187	.8106	.8025	.7945	.7866	.7788	.7711	.7634	.7558	.7483
.3	.7408	.7334	.7261	.7189	.7118	.7047	.6977	.6907	.6839	.6771
.4	.6703	.6636	.6570	.6505	.6440	.6376	.6313	.6250	.6188	.6126
.5	.6065	.6005	.5945	.5886	.5827	.5770	.5712	.5655	.5599	.5543
.6	.5488	.5434	.5379	.5326	.5273	.5220	.5169	.5117	.5066	.5016
.7	.4966	.4916	.4868	.4819	.4771	.4724	.4677	.4630	.4584	.4538
.8	.4493	.4449	.4404	.4360	.4317	.4274	.4232	.4190	.4148	.4107
.9	.4066	.4025	.3985	.3946	.3906	.3867	.3829	.3791	.3753	.3716
1.0	.3679	.3642	.3606	.3570	.3535	.3499	.3465	.3430	.3396	.3362
1.1	.3329	.3296	.3263	.3230	.3198	.3166	.3135	.3104	.3073	.3042
1.2	.3012	.2892	.2952	.2923	.2894	.2865	.2837	.2808	.2780	.2753
1.3	.2725	.2698	.2671	.2645	.2618	.2592	.2567	.2541	.2516	.2491
1.4	.2466	.2441	.2417	.2393	.2369	.2346	.2322	.2299	.2276	.2254
1.5	.2231	.2209	.2187	.2165	.2144	.2122	.2101	.2080	.2060	.2039
1.6	.2019	.1999	.1979	.1959	.1940	.1920	.1901	.1882	.1864	.1845
1.7	.1827	.1809	.1791	.1773	.1755	.1738	.1720	.1703	.1686	.1670
1.8	.1653	.1637	.1620	.1604	.1588	.1572	.1557	.1541	.1526	.1511
1.9	.1496	.1481	.1466	.1451	.1437	.1423	.1409	.1395	.1381	.1367
2.0	.1353	.1340	.1327	.1313	.1300	.1287	.1275	.1262	.1249	.1237
2.1	.1225	.1212	.1200	.1188	.1177	.1165	.1153	.1142	.1130	.1119
2.2	.1108	.1097	.1086	.1075	.1065	.1054	.1044	.1033	.1023	.1013
2.3	.1003	.0993	.0983	.0973	.0963	.0954	.0944	.0935	.0926	.0916
2.4	.0907	.0898	.0889	.0880	.0872	.0863	.0854	.0846	.0837	.0829
2.5	.0821	.0813	.0805	.0797	.0789	.0781	.0773	.0765	.0758	.0750
2.6	.0743	.0735	.0728	.0721	.0714	.0707	.0699	.0693	.0686	.0679
2.7	.0672	.0665	.0659	.0652	.0646	.0639	.0633	.0627	.0620	.0614
2.8	.0608	.0602	.0596	.0590	.0584	.0578	.0573	.0567	.0561	.0556
2.9	.0550	.0545	.0539	.0534	.0529	.0523	.0518	.0513	.0508	.0503
3.0	.0498	.0493	.0488	.0483	.0478	.0474	.0469	.0464	.0460	.0455
3.1	.0450	.0446	.0442	.0437	.0433	.0429	.0424	.0420	.0416	.0412
3.2	.0408	.0404	.0400	.0396	.0392	.0388	.0384	.0380	.0376	.0373
3.3	.0369	.0365	.0362	.0358	.0354	.0351	.0347	.0344	.0340	.0337
3.4	.0334	.0330	.0327	.0324	.0321	.0317	.0314	.0311	.0308	.0305
3.5	.0302	.0299	.0296	.0293	.0290	.0287	.0284	.0282	.0279	.0276
3.6	.0273	.0271	.0268	.0265	.0263	.0260	.0257	.0255	.0252	.0250
3.7	.0247	.0245	.0242	.0240	.0238	.0235	.0233	.0231	.0228	.0226
3.8	.0224	.0221	.0219	.0217	.0215	.0213	.0211	.0209	.0207	.0204
3.9	.0202	.0200	.0198	.0196	.0194	.0193	.0191	.0189	.0187	.0185
4.0	.0183	.0181	.0180	.0178	.0176	.0174	.0172	.0171	.0169	.0167
4.1	.0166	.0164	.0162	.0161	.0159	.0158	.0156	.0155	.0153	.0151
4.2	.0150	.0148	.0147	.0146	.0144	.0143	.0141	.0140	.0138	.0137
4.3	.0136	.0134	.0133	.0132	.0130	.0129	.0128	.0127	.0125	.0124
4.4	.0123	.0122	.0120	.0119	.0118	.0117	.0116	.0114	.0113	.0112
4.5	.0111	.0110	.0109	.0108	.0107	.0106	.0105	.0104	.0103	.0102
4.6	.0101	.0100	.0099	.0098	.0097	.0096	.0095	.0094	.0093	.0092
4.7	.0091	.0090	.0089	.0088	.0087	.0087	.0086	.0085	.0084	.0083
4.8	.0082	.0081	.0081	.0080	.0079	.0078	.0078	.0077	.0076	.0075
4.9	.0074	.0074	.0073	.0072	.0072	.0071	.0070	.0069	.0069	.0068
5.0	.0067									

TABLE 3 OPTIMUM VALUE OF ρ FOR M/M/I/N SYSTEMS

Where E = average cost per service G = average profit per service N = maximum system size.

E/G \ N	1	2	3	4	5	6	7	8	9	10	12	14	16	18	20
0.05	0.29	0.36	0.44	0.50	0.55	0.59	0.63	0.66	0.68	0.70	0.74	0.77	0.79	0.81	0.82
0.10	0.46	0.50	0.55	0.60	0.64	0.68	0.70	0.73	0.75	0.77	0.80	0.82	0.84	0.85	0.87
0.15	0.63	0.61	0.64	0.68	0.72	0.74	0.77	0.79	0.80	0.81	0.84	0.86	0.87	0.88	0.89
0.20	0.81	0.71	0.73	0.75	0.77	0.80	0.81	0.83	0.84	0.85	0.87	0.88	0.90	0.91	0.91
0.25	1.00	0.82	0.81	0.82	0.83	0.84	0.85	0.87	0.87	0.88	0.90	0.91	0.92	0.93	0.93
0.30	1.21	0.93	0.88	0.88	0.88	0.89	0.89	0.90	0.90	0.91	0.92	0.93	0.94	0.95	0.95
0.35	1.45	1.03	0.96	0.94	0.93	0.93	0.93	0.94	0.94	0.94	0.95	0.95	0.96	0.96	0.96
0.40	1.72	1.16	1.04	1.00	0.98	0.97	0.97	0.97	0.97	0.97	0.97	0.97	0.98	0.98	0.98
0.45		1.28	1.12	1.06	1.04	1.02	1.01	1.00	1.00	1.00	0.99	0.99	0.99	0.99	0.99
0.50		1.43	1.21	1.13	1.09	1.06	1.05	1.04	1.03	1.02	1.02	1.01	1.01	1.01	1.01
0.55		1.59	1.31	1.20	1.15	1.11	1.09	1.07	1.06	1.05	1.04	1.03	1.03	1.02	1.02
0.60		1.77	1.42	1.28	1.21	1.16	1.13	1.11	1.10	1.09	1.07	1.05	1.05	1.04	1.04
0.65		1.98	1.55	1.37	1.27	1.22	1.18	1.15	1.13	1.12	1.09	1.08	1.07	1.06	1.05
0.70			1.69	1.47	1.35	1.28	1.24	1.20	1.18	1.15	1.12	1.10	1.09	1.08	1.07
0.75			1.86	1.58	1.44	1.35	1.30	1.25	1.22	1.19	1.15	1.13	1.11	1.11	1.09
0.80				1.73	1.55	1.43	1.36	1.31	1.28	1.24	1.20	1.16	1.14	1.12	1.11
0.85				1.93	1.69	1.55	1.46	1.39	1.35	1.30	1.24	1.20	1.16	1.16	1.14
0.90					1.89	1.70	1.58	1.50	1.44	1.38	1.31	1.26	1.22	1.19	1.17
0.95						1.98	1.79	1.64	1.55	1.51	1.41	1.35	1.29	1.26	1.23
0.99										1.80	1.65	1.50	1.45	1.40	1.35

REFERENCES

For an introduction to the theory of random arrival, random service queueing systems (M/M/-/-), readers are referred to the following general introductory textbooks on Operational Research.

Sasieni, M.W., et al., Operations Research (Wiley, New York, 1959) pp. 125-54.

Wagner, H.M., Principles of Operational Research (Prentice-Hall, Englewood Cliffs, N.J., 1969) pp. 837-85.

Houlden, B.T., Some Techniques of Operational Research (English Universities Press, London, 1969) pp. 75-115.

Churchman, C.W., Ackoff, R.L., and Arnoff, E.L., Introduction to Operations Research (Wiley, New York, 1957) pp. 391-416.

Specially recommended for readers wishing to study Queueing Theory in greater depth is

Page, E.G., Queueing Theory in O.R. (Butterworth, London, 1972).

Other specialistic textbooks on Queueing Theory include

Cohen, J.W., The Single Server Queue (North-Holland, Amsterdam, 1969).

Cox, D.R., and Smith, W.L., Queues (Chapman & Hall, London, 1971).

Fry, R.C., Probability and Engineering Uses (Van Nostrand, New York, 1928). (For proof of M/D/1 formulae, see p. 374).

Lee, A.M., Applied Queueing Theory (Macmillan, London and Basingstoke, 1966).

Jaiswal, N.K., Priority Queues (Academic Press, London, 1968).

Khintchine, A.Y., Mathematical Models in the Theory of Queueing (Hafner, New York, 1968).

Morse, P.M., Queues, Inventories and Maintenance (Wiley, New York, 1958).

Saaty, R.L., Elements of Queueing Theory (McGraw-Hill, London, 1961).